# MANUAL OF
# ACTIVE FILTER DESIGN

# MANUAL OF
# ACTIVE FILTER DESIGN

## JOHN L. HILBURN

*President, Microcomputer Systems, Inc.*

## DAVID E. JOHNSON

*Department of Electrical Engineering,
Louisiana State University*

## Second Edition

**McGraw-Hill Book Company**

New York   St. Louis   San Francisco   Auckland   Bogotá
Hamburg   Johannesburg   London   Madrid   Mexico
Montreal   New Delhi   Panama   Paris   São Paulo
Singapore   Sydney   Tokyo   Toronto

Library of Congress Cataloging in Publication Data

Hilburn, John L. (date)
  Manual of active filter design.

  Bibliography: p.
  Includes index.
    1. Electric filters, Active.   2. Electronic
circuit design—Graphic methods.   I. Johnson,
David E.   II. Title.
TK7872.F5H5   1983      621.3815′324      82-4649
ISBN 0-07-028769-4

1234567890   DOC/DOC   898765432

ISBN 0-07-028769-4

*The editors for this book were Harry Helms and Peggy Lamb,
and the production supervisor was Thomas G. Kowalczyk.
It was set in Caledonia by University Graphics, Inc.*

*Printed and bound by R. R. Donnelley & Sons.*

# CONTENTS

# LIST OF DESIGN GRAPHS

# PREFACE

In this book we present simplified methods for obtaining a complete, practical filter design by inspection of a graph, requiring no computations whatsoever. The book therefore is useful to all filter designers, from the novice to the expert. The filter circuit elements used are operational amplifiers and standard values of resistances and capacitances.

The type of filters which one may construct using the graphs are the following:

1. Low-pass (Butterworth or Chebyshev of second or fourth orders)
2. High-pass (Butterworth or Chebyshev of second or fourth orders)
3. Band-pass (second and higher orders)
4. Band-reject (notch)
5. Phase-shift or delay (all-pass or Bessel)

Each of the filter types is discussed in a separate chapter. At the end of each chapter the design procedure is summarized and the appropriate graphs are presented. Practical design suggestions are given for each circuit considered.

Examples are given of every type of filter considered and actual photographs of the results are included. A detailed example is presented in Sec. 2.3, which may be used as a design guideline. However, it is not necessary to read the chapters in order to use the handbook, since all the necessary information is presented on the summary pages of each chapter.

In this second edition we have retained the popular filter designs that have been widely used and well accepted in the first edition. We have added, in the cases of the low-pass and high-pass filters, the very useful 0.1 dB Chebyshev designs. The other major addition is the inclusion, in the cases of low-pass and band-pass filters, of the important biquad filter designs, which have superior performance characteristics and are extremely easy to tune. It is probably no exaggeration to say that the biquad is the best of all possible filter designs available.

*John L. Hilburn*
*David E. Johnson*

# 1

# INTRODUCTION

## 1.1 Active Filters

A filter is a device which passes signals of certain frequencies and rejects or attenuates those of other frequencies. Passive filters are constructed with inductors, capacitors, and resistors, but for certain frequency ranges inductors, because of their size and practical performance limitations, are undesirable. Consequently there has been, for some years, a trend toward replacing inductors by active devices which simulate the effect of inductors. This trend has accelerated with the advances in miniaturization which have made the active devices available at prices competitive with, and in many cases cheaper than, those of inductors.

In this manual we present simplified methods for constructing a variety of active filters having specified characteristics with standard element values. The active device we use is the integrated circuit (IC) operational amplifier, which is briefly described in the remainder of this chapter. Graphs are presented for each type of filter, and, depending on the specifications, the designer may simply choose the appropriate graph and read off the circuit element values. For the designer interested in the theoretical details, there is a chapter of background material with numerous references provided for each filter type. However, to use the manual, one needs only to refer to the summary sheet preceding the graphs at the end of each chapter.

## 1.2 The Operational Amplifier

The basic element we shall use in the construction of an active filter is the operational amplifier (op-amp), the symbol for which is shown in Fig. 1.1. Only three terminals are shown in the figure, the inverting input terminal (−), the noninverting input terminal (+), and the output terminal. How-

**1**

ever, a practical op-amp is actually a multiterminal device. The purposes of the other terminals are specified by the manufacturer and include, in general, power supply connections, frequency compensation terminals, and offset null terminals.

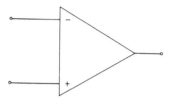

**Fig. 1.1    A differential op-amp.**

The equations we have derived in the following chapters are obtained assuming zero voltage between the two input terminals of the op-amp and zero current into the two input terminals. This is true of the ideal op-amp, and is closely approximated by practical op-amps, if used according to the manufacturer's specifications.

Numerous publications are available describing in detail the characteristics and uses of commercially available operational amplifiers. (See, for example [1]–[14].)° In addition, most manufacturers publish detailed catalogs containing information on their specific op-amps. An extensive list of manufacturers is given in [1]. Some well-known manufacturers include Burr-Brown Research Corp., Fairchild Semiconductor, Motorola, National Semiconductor, RCA, Signetics Corp., and Texas Instruments.

The op-amp of Fig. 1.1 is a differential-input amplifier, which is a commonly manufactured type. In general, for stable operation, IC op-amps require frequency compensation. Some, such as the 741, are internally compensated. The $\mu$A741, AD741, MC1741, LM741, CA741, etc. are all type 741 op-amps. The different representations are used to identify the manufacturer. Other types of compensated op-amps include the 536, 107, 5556, 740, and 747 (dual 741). Other op-amps require external compensation, specified by the manufacturer, but generally are useful for much higher frequencies and gains. Some examples of these are the types 709, 748, 101, and 531.

For best results in the circuit configurations given in the following chapters, the designer should use op-amps which perform adequately for the

°References thus cited are listed in the Bibliography.

gains and frequency ranges of interest. For example, the open-loop gains as specified by the manufacturer should be at least 50 times the filter gain [4]. Other suggestions will be made on the summary sheets at the end of each chapter.

## 1.3   Resistors and Capacitors

There are three types of resistors in common use. The carbon composition resistor is the most widely used and is acceptable in most noncritical filter applications. This is particularly true if the filter is used at room temperature. In all our examples the filters were constructed with 5% tolerance carbon composition resistors. These were used because they are the most economical and commonly available. For high-performance applications, or in instances where temperature is important, one should use either metal-film or wire-wound resistors.

In the case of capacitors, the ceramic disk capacitor is a very common and economical type. However, these should be used in the most noncritical applications. A more acceptable common type is the Mylar capacitor, which is the type we used in most of our examples. For critical applications and high performance, polystyrene and Teflon capacitors are good choices in most cases.

For a good discussion of resistors and capacitors, the reader is referred to [2], pp. 317–319.

## 1.4   Basic Op-Amp Circuits

The filters that are designed in the book perform best when the input signals are from low-output-impedance sources. If the signal that is to be filtered is not of this type, it may be desirable to *precondition*, or *preamplify*, the signal with an op-amp circuit as shown in Fig. 1.2. The circuit consists of an op-amp and two resistors $R_a$ and $R_b$, and has the property that its output voltage $V_1$ is related to its input voltage $V_{in}$ by

$$V_1 = \mu V_{in} \tag{1.1}$$

where

$$\mu = 1 + \frac{R_a}{R_b} \tag{1.2}$$

The circuit is called a *voltage-controlled voltage source* (VCVS) and $\mu$ is its *gain*.

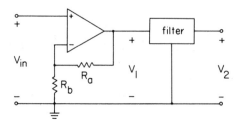

**Fig. 1.2   A VCVS circuit cascaded with a filter circuit.**

In the case of Fig. 1.2, the VCVS is cascaded with a filter circuit with input voltage $V_1$ (the output of the VCVS) and output voltage $V_2$. The advantages of this arrangement are that the VCVS approximates the zero-output impedance we need and, by adjusting the gain $\mu$, we may obtain a predetermined level of the filter output voltage $V_2$. The filter itself is designed to provide certain predetermined levels, or *gains*, but with the added flexibility of adjusting $\mu$ we may restrict the filter designs to a relatively small number of filter gains.

As an example, if the filter of Fig. 1.2 is designed to give at a certain frequency,

$$V_2 = 2V_1$$

and we wish to have

$$V_2 = 6V_{in}$$

then we need

$$V_1 = 3V_{in}$$

Thus the VCVS must have a gain of

$$\mu = 1 + \frac{R_a}{R_b} = 3$$

or

$$R_a = 2R_b$$

If we arbitrarily select $R_b = 10$ kΩ, then we require $R_a = 20$ kΩ.

If in the VCVS of Fig. 1.2 we make $R_a = 0$ (a short circuit) and $R_b$ infinite (an open circuit), then we have the circuit of Fig. 1.3. By Eq. (1.2) its gain is $\mu = 1$, so that

$$V_1 = V_{in} \tag{1.3}$$

Thus the output voltage $V_1$ is the same as the input voltage $V_{in}$, but there is the important difference that $V_1$ is the output of a zero-output-impedance device whereas $V_{in}$ may be any voltage. In addition the device draws no current because of the presence of the op-amp, and thus it acts as a *buffer* stage between $V_{in}$ and $V_1$. For this reason it is referred to as a *buffer amplifier*, or, because the output $V_1$ *follows* the input $V_{in}$, as a *voltage follower*. The buffer amplifier is useful when we need to precondition the filter input signal but all the gain we need is provided by the filter itself.

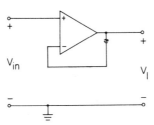

**Fig. 1.3    A buffer amplifier.**

Another popular op-amp circuit is that of Fig. 1.4, for which

$$V_{out} = -\frac{R_a}{R_b} V_{in} \tag{1.4}$$

This circuit thus is a nearly zero-output-impedance device that provides a gain of $R_a/R_b$ and *inverts* (changes the sign of) the input signal. For this reason it is called an *inverting amplifier*, or simply an *inverter*.

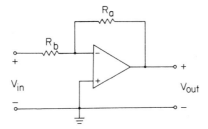

**Fig. 1.4    An inverter.**

The VCVS, buffer amplifier, and inverter circuits thus are useful in preconditioning the input signals when necessary. As we will see, they are also very popular components of many filter circuits.

# 2 LOW-PASS FILTERS

## 2.1 General Circuits and Equations

A low-pass filter is a device which passes signals of low frequencies and suppresses or attenuates those of high frequencies. Its performance may be illustrated by its amplitude response, which is a plot of the amplitude $|H(j\omega)|$ of its transfer function $H(s)$ versus frequency $\omega$ (radians/sec) or $f$ (Hz), where $\omega = 2\pi f$. In all cases we shall take $H(s) = V_2(s)/V_1(s)$, where $V_2$ is the output voltage and $V_1$ is the input voltage. An example is shown in Fig. 2.1, where the response represented by the broken line is the ideal case and the response represented by the solid line is a realizable approximation of the ideal. The value $\omega_c$ (or in Hz, $f_c = \omega_c/2\pi$) is the cutoff frequency defined as the point at which $|H(j\omega)|$ is $1/\sqrt{2} = 0.707$ times its maximum value, shown here as $A$. The passband is the range $0 \leq \omega < \omega_c$ and the stopband is the range $\omega > \omega_c$.

Alternatively the amplitude response may be plotted as amplitude in decibels (dB), which we denote by $\alpha$, versus frequency $\omega$ (or $f$), or versus

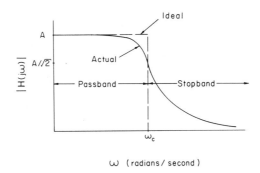

$\omega$ (radians / second)

**Fig. 2.1 Low-pass amplitude response.**

$\log \omega$ (or $\log f$). An example is shown in Fig. 2.2, where it may be seen that cutoff corresponds to $\alpha = -3$ dB.

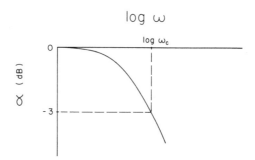

**Fig. 2.2    Amplitude response in dB.**

A second-order approximation to an ideal low-pass filter is achieved by the transfer function

$$\frac{V_2(s)}{V_1(s)} = \frac{K}{s^2 + as + b} \tag{2.1}$$

where $a$ and $b$ are properly chosen constants and $K$ is a constant [15]. The term *second-order* refers to the degree of the denominator polynomial. Higher order transfer functions are like Eq. (2.1) except that the denominator is of higher degree. The gain of the low-pass filter is the value of its transfer function at $s = 0$, and is given in the case of Eq. (2.1) by gain = $K/b$.

There are any number of ways of obtaining low-pass filters using active devices in lieu of inductors. (See, for example [2], [6], [15], [16].) One method we use is that of Sallen and Key [17], in which the active device is an operational amplifier (op-amp), described in Chapter 1. A Sallen and Key second-order low-pass filter is shown in Fig. 2.3, where the resistors and capacitors are properly chosen to realize given values of $a$ and $b$ in Eq. (2.1). The op-amp, together with the resistors $R_3$ and $R_4$, constitutes a voltage-controlled voltage source (VCVS), and hence the Sallen and Key network is of the VCVS type.

Higher order filters may be obtained by cascading two or more second-order filters. For example, a fourth-order low-pass filter is obtained by cascading two networks such as that of Fig. 2.3, as we shall see in Sec. 2.6.

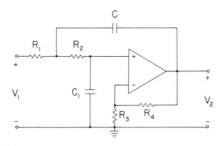

**Fig. 2.3  A second-order VCVS low-pass filter.**

Analysis of Fig. 2.3 shows that it achieves Eq. (2.1) with

$$K = \frac{\mu}{R_1 R_2 C C_1}$$

$$a = \frac{1}{R_2 C_1}(1 - \mu) + \frac{1}{R_1 C} + \frac{1}{R_2 C} \qquad (2.2)$$

$$b = \frac{1}{R_1 R_2 C C_1}$$

where

$$\mu = 1 + \frac{R_4}{R_3}$$

The quantity $\mu$ is the gain of the VCVS, and is also the gain of the filter since $K/b = \mu$.

Another low-pass filter circuit that we consider is the so-called *biquad* circuit [18] of Fig. 2.4. It has more elements than the VCVS circuit, but it is much easier to adjust, or *tune*, and is much more stable, particularly in higher order cases. Analysis of Fig. 2.4 shows that Eq. (2.1) is achieved with

$$K = \frac{1}{R_1 R_4 C^2}$$

$$a = \frac{1}{R_2 C} \qquad (2.3)$$

$$b = \frac{1}{R_3 R_4 C^2}$$

The gain $K/b$ is therefore $R_3/R_1$.

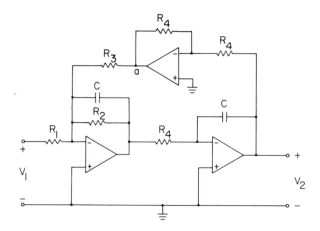

**Fig. 2.4   A second-order biquad low-pass filter.**

The ease of adjustment of the biquad may be seen from these results. If we select $C$ and $R_4$ arbitrarily, then the value of $K$ is determined solely by $R_1$, $a$ is determined solely by $R_2$, and $b$ is determined solely by $R_3$. In addition, the presence of the inverter composed of the op-amp and its two connected resistors, both labeled $R_4$, allows us to obtain an inverted output $-V_2$ at node $a$.

There are many types of low-pass filters, but the two most commonly used are the Butterworth and Chebyshev types. In the following sections we shall give a brief description of these and present a simple technique for their design.

## 2.2   Low-Pass Butterworth Filters

A filter which approximates the ideal low-pass filter with a relatively flat passband characteristic is the Butterworth filter [19–22]. Its amplitude response is given by

$$|H(j\omega)| = \frac{K}{\sqrt{1 + (\omega/\omega_c)^{2n}}} \tag{2.4}$$

where $n$ is the order of the filter. Some examples are shown in Fig. 2.5, where it may be seen that the filter improves as $n$ increases. We shall consider in Sec. 2.3 the case $n = 2$, and the case $n = 4$ in Sec. 2.6.

The Butterworth filter has the advantage of a so-called maximally flat, monotonic response in the passband. However, its cutoff characteristics are inferior to those of the Chebyshev filter, which we consider in Sec. 2.4. For example, the amplitudes shown in Fig. 2.5 are attenuated, for relatively large frequencies, at an approximate rate of $-20n$ dB/decade [19], which is the same as that of the Chebyshev filter. (A decade is the frequency interval between two frequencies, one being 10 times the other.) However, the attenuation in the stopband is considerably more for the Chebyshev than for the Butterworth filter.

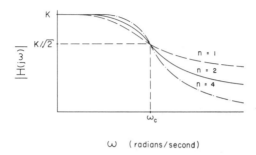

**Fig. 2.5   Butterworth amplitude responses.**

## 2.3   Second-Order Low-Pass Butterworth Filters

To construct a low-pass VCVS second-order Butterworth filter, we use Fig. 2.3 and determine practical values of the capacitances $C$ and $C_1$, and the resistances $R_1$, $R_2$, $R_3$, and $R_4$ so that Eqs. (2.2) are satisfied for a given cutoff frequency $f_c$.

To facilitate the development, we have constructed a number of graphs which may be used to obtain the capacitances and resistances by inspection, as follows. We locate the given $f_c$ on either Fig. 2.12a, b, or c. For $f_c$ between 1 Hz and $10^2 = 100$ Hz, we use Fig. 2.12a; between 100 Hz and $10^4 = 10,000$ Hz, we use Fig. 2.12b; and between 10,000 Hz and $10^6 = 1,000,000$ Hz, we use Fig. 2.12c. Once we have located $f_c$ we have a variety of standard values shown on the figure from which to choose the capacitance $C$. This choice of $C$ then determines from the figure a value of a $K$ parameter. We are now ready to find the capacitance $C_1$ and the values of the resistances. Figures 2.13 through 2.17 are used for this purpose for the second-order low-pass Butterworth filter. If we want a gain of 2 for the filter, we use Fig. 2.13, which yields for the given $K$ parameter the values of the resistances and the capacitance $C_1$, which in this case is the same as $C$. For gains of 4, 6, 8, or 10, we use Figs. 2.14 through 2.17, as indicated, to find the resistances and $C_1$, which will either be $C$, or for the higher gains, $2C$, as shown. (We have not included gains larger than 10 because of the large spread in element values and high sensitivities in these cases [2].) Finally we select standard resistance values as close as possible to those indicated by the graph and build the circuit.

In summary, the curves allow one to choose $f_c$ between 1 and 1,000,000 Hz, a standard value of $C$, and a gain of 2, 4, 6, 8, or 10, and construct a second-order low-pass Butterworth filter as displayed in Fig. 2.3, with a standard value of $C_1$ and practical values of resistances. We may add that if one wishes to change the cutoff point from one value to another, one may do so by changing only the capacitors in the circuit by simply selecting capacitor values for the new cutoff point which do not change the $K$ parameter in Fig. 2.12$a$, $b$, or $c$.

As an example, suppose we want $f_c$ to be 1000 Hz. From Fig. 2.12$b$ we see that we may select $C$ from a range of 0.001 to 0.1 microfarads ($\mu$F). Suppose we have available a 0.01 $\mu$F capacitor which we choose as $C$. From the figure this determines a $K$ parameter of 10. Suppose also we desire a gain of 6 for the filter network. Then we use Fig. 2.15, which yields for a $K$ parameter of 10, the values $R_1 = 6.2$ k$\Omega$, $R_2 = 20.7$ k$\Omega$, $R_3 = 32$ k$\Omega$, $R_4 = 160$ k$\Omega$, and $C_1 = 2C = 0.02$ $\mu$F. Standard values of resistances very near the calculated values are 6.2, 20, 33, and 160 k$\Omega$. The response of a filter utilizing these standard values of 5% tolerances with a $\mu$A741 op-amp is shown in Fig. 2.6. The actual results are $f_c = 976$ Hz with a gain of 5.9. In the picture the amplitude response starts at 0 Hz, and each division represents 500 Hz.

Practical suggestions concerning the circuit elements are given in the summary of second-order VCVS filters which follows this chapter.

Finally we should note that the curves in the figures are plotted in logarithmic scale and care should be exercised in interpreting them. For instance, if in the foregoing example we choose $C = 0.002$ $\mu$F, from Fig. 2.12$b$ we have $K = 50$. The small 5 between 10 and $10^2 = 100$ means 50. If $C = 0.03$ $\mu$F is chosen, then $K = 3.35$. Finally if, instead of 1000 Hz, $f_c = 2000$ Hz, which is located at the small 2 between $10^3$ and $10^4$ on the horizontal axis, then a value $C = 0.003$ $\mu$F yields $K = 16.3$.

To construct the low-pass biquad circuit of Fig. 2.4, the capacitance $C$ and the frequency $f_c$ are used as before to determine the $K$ parameter from the same curves, namely Fig. 2.12$a$, $b$, or $c$. The resistances $R_1$, $R_2$, $R_3$, and $R_4$ are then found from Fig. 2.33, with $R_2$, $R_3$, and $R_4$ read directly from the graph, and $R_1$ obtained from

$$R_1 = \frac{R_3}{\text{GAIN}}$$

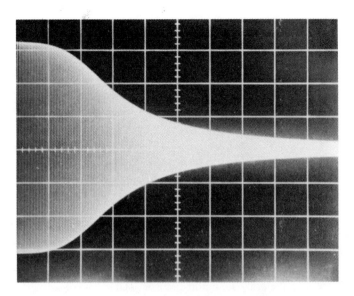

**Fig. 2.6   A second-order low-pass Butterworth response.**

In the previous example, using $C = 0.01\ \mu F$ for the two capacitors and the frequency of $f_c = 1000$ Hz, we have from Fig. 2.12$b$ a $K$ parameter of 10. From Fig. 2.33 we have $R_2 = 11.3$ k$\Omega$, $R_3 = 16$ k$\Omega$, and $R_4 = 16$ k$\Omega$. For the gain of 6 we have

$$R_1 = \frac{R_3}{6} = \frac{16}{6} = 2.7\ k\Omega$$

Practical suggestions for the biquad filter are given in the summary at the end of the chapter.

### 2.4   Low-Pass Chebyshev Filters

The low-pass Chebyshev filter has the amplitude response [20]

$$|H(j\omega)| = \frac{K}{\sqrt{1 + \epsilon^2\ C_n^2(\omega/\omega_c)}} \tag{2.5}$$

where $\epsilon$ is a constant and $C_n$ is the Chebyshev polynomial of the first kind of degree $n$. The response exhibits ripples in the passband, the number of which depends on $n$, as shown in Fig. 2.7.

The width of the ripple is determined by $\epsilon$ and may be used to charac-

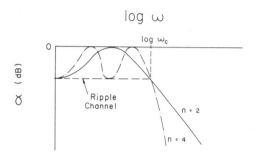

Fig. 2.7   Chebyshev filter response.

terize the filter. For example, a ½ dB Chebyshev low-pass filter is one with a response whose ripple width is ½ dB. Thus $\omega_c$ (or $f_c$) is the cutoff frequency in the sense that it is the terminal frequency of the ripple channel. In the case of a 3 dB Chebyshev filter, $\omega_c$ is also the cutoff point in the conventional sense. In general, the cutoff point is related to $\omega_c$ as used here, and the specific relationship is given in Sec. 2.5 for the second-order case, and in Sec. 2.6 for the fourth-order case.

The Chebyshev filter has the disadvantage, as compared with the Butterworth filter, of ripples occurring in the passband. However, it is the best of all filters of the type defined in Eq. (2.1) from the standpoint of its cutoff characteristics [23]. For example, $\alpha$ in the Chebyshev case is approximately $6(n - 1) + 20 \log \epsilon$ below the Butterworth filter in dB outside the passband [19].

### 2.5   Second-Order Low-Pass Chebyshev Filters

Since the only difference between the second-order Chebyshev and Butterworth filters is the value of $a$ and $b$ in Eq. (2.1), the technique for designing a Chebyshev filter of the form of Figs. 2.3 or 2.4 is identical to that of the Butterworth filter described in detail in Sec. 2.3. The $K$ parameter is determined from the same curves, namely Fig. 2.12$a$, $b$, or $c$, as in the Butterworth case, and the other parameters are obtained from the appropriate one of Figs. 2.18 through 2.32 and 2.34 through 2.38, depending on the gain and the ripple width desired.

As was observed in Sec. 2.4, in the low-pass Chebyshev filter, $f_c$ is the length of the ripple channel, rather than the conventional cutoff point. For the second-order case the conventional cutoff point in Hz is 1.94 $f_c$ for the

$\%_0$ dB filter, 1.39 $f_c$ for the $\%$ dB filter, 1.21 $f_c$ for the 1 dB, 1.07 $f_c$ for the 2 dB, and $f_c$ for the 3 dB.

As an example, suppose we require $f_c$ = 2750 Hz and we desire a 1 dB VCVS filter with a gain of 2. From Fig. 2.12$b$, selecting $C$ = 0.0047 $\mu$F yields a $K$ parameter of 7.8. From Fig. 2.24, which corresponds to the 1 dB ripple case, we find for this value of $K$, $R_1$ = 11.3 k$\Omega$, $R_2$ = 12.3 k$\Omega$, $R_3$ = $R_4$ = 48 k$\Omega$, and $C_1$ = $C$ = .0047 $\mu$F. Practical values of resistance approximating the calculated values are 11, 12, and 47 k$\Omega$. Using these values and a $\mu$A741 op-amp, we have the resulting response shown in Fig. 2.8. The cutoff frequency (at the termination of the ripple band) was 2841 Hz, the gain was 2, and the ripple width was 1.16 dB. The scale used in the picture was 500 Hz/division.

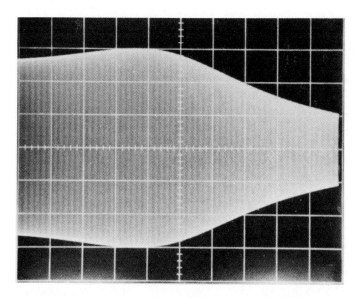

**Fig. 2.8   A second-order low-pass Chebyshev response.**

## 2.6   Fourth-Order Low-Pass Butterworth and Chebyshev Filters

We may obtain fourth-order VCVS low-pass Butterworth or Chebyshev filters by cascading two networks like that of Fig. 2.3, as shown in Fig. 2.9. The procedure for finding practical values of resistances and capacitances is identical to that described for the second-order Butterworth filter in Sec. 2.3, except, of course, that there are more parameters.

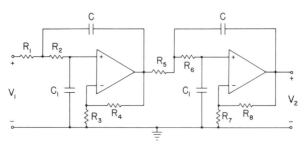

**Fig. 2.9    A fourth-order VCVS low-pass filter.**

We shall illustrate the procedure by obtaining a fourth-order 3 dB Chebyshev low-pass filter with $f_c$ = 1000 Hz and a gain of 36. From Fig. 2.12$b$, we select $C$ = 0.01 $\mu$F, resulting in a $K$ parameter of 10. From Fig. 2.55, for this value of $K$, we have $R_1$ = 9.3 k$\Omega$, $R_2$ = 15 k$\Omega$, $R_3$ = 29 k$\Omega$, $R_4$ = 145 k$\Omega$, $R_5$ = 16 k$\Omega$, $R_6$ = 41 k$\Omega$, $R_7$ = 67 k$\Omega$, $R_8$ = 330 k$\Omega$, and $C_1$ = $2C$ = 0.02 $\mu$F. Using standard resistances of 9.1, 15, 30, 150, 16, 39, 68, and 330 k$\Omega$ and a $\mu$A747 op-amp, we obtain the characteristics shown in Fig. 2.10. The scale shown is 200 Hz/division, and the actual results are $f_c$ = 972 Hz, 3 dB, and a gain of 36.8.

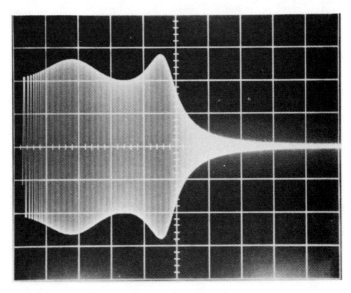

**Fig. 2.10    A fourth-order low-pass Chebyshev response.**

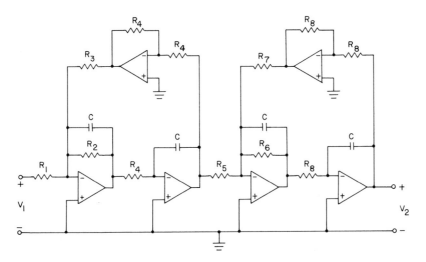

**Fig. 2.11   A fourth-order biquad low-pass filter.**

Fourth-order biquad circuits are obtained by cascading two biquads as shown in Fig. 2.11. For the given $f_c$ and selected value of $C$, the $K$ parameter is determined from the appropriate one of Fig. 2.12a, b, or c. The resistances are then found from the appropriate one of Figs. 2.57 to 2.62. Each of the two stages has been designed to have the square root of the gain. Thus the circuit has a gain of

$$\sqrt{\text{GAIN}} \ \sqrt{\text{GAIN}} = \text{GAIN}$$

If the designer wishes, the stage gains may be assigned differently, as long as their product is the filter gain. For example, if the gain is 4, the graphs give each stage a gain of $\sqrt{4} = 2$. We could choose other values, like gain = 1 for the first stage and gain = 4 for the second stage.

For the convenience of the designer, all the techniques described in this chapter for constructing low-pass filters are summarized following this section. The second-order filters together with their graphs are presented first, followed by the fourth-order filters and their graphs.

## SUMMARY OF LOW-PASS SECOND-ORDER VCVS FILTER DESIGN PROCEDURE

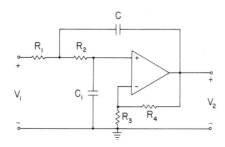

**General circuit.**

### Procedure

Given $f_c$ (Hz), gain, and filter type (Butterworth or Chebyshev), perform the following steps:

1. Select a value of capacitance $C$, determining a $K$ parameter from Fig. 2.12$a$ if $f_c$ is between 1 and $10^2 = 100$, from Fig. 2.12$b$ if $f_c$ is between 100 and $10^4 = 10,000$, and from Fig. 2.12$c$ if $f_c$ is between 10,000 and $10^6 = 1,000,000$ Hz.

2. Using this value of $K$, find the remaining element values of the circuit from the appropriate one of Figs. 2.13 through 2.17 for the Butterworth filter, and Figs. 2.18 through 2.32 for the Chebyshev filter, depending on the gain and, in the Chebyshev case, the dB ripple desired.

3. Select standard element values which are as close as possible to those indicated on the graph and construct the circuit.

### Comments and Suggestions

The curves are designed for 35 standard values of capacitance. Any intermediate values of capacitance may be used by observing that dividing the capacitance values by a constant $k$ multiplies the cutoff frequency $f_c$ by $k$. The resistances remain unchanged. This procedure of changing the capacitances may be accomplished by interpolation on the frequency versus $K$ parameter graphs.

If the op-amp to be used has a low-input resistance (less than 250 k$\Omega$), values of $K$ from 1 to 10 give best results. For higher input resistances (like 1 MΩ), $K$ values up to 25 are acceptable, and for very high input resistances, such as those associated with field-effect transistor (FET) op-amps, values of $K$ up to 100 may be used in most cases.

The values on the graphs for $R_3$ and $R_4$ were determined to minimize the dc offset of the op-amp. Other values of $R_3$ and $R_4$ may be used as long as the ratio $R_4/R_3$ is the same as that of the graph values. Standard element values of 5% tolerance normally yield acceptable results, but for best performance higher precision elements with values close to the graph values should be used. This is especially true for the higher gains where the element values are much more critical.

Finally there must be a dc return to ground at the filter input, the open-loop gain of the op-amp should be at least 50 times the gain of the filter at $f_c$, and the desired peak-to-peak voltage at $f_c$ should not exceed $10^6/\pi f_c$ times the slew rate of the op-amp.

A specific example of a second-order design is given in Sec. 2.3.

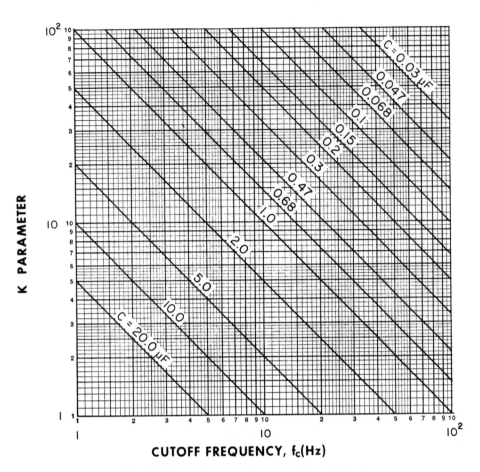

**K PARAMETER**

**CUTOFF FREQUENCY, $f_c$(Hz)**

Fig. 2.12(*a*)   *K* parameter versus frequency.

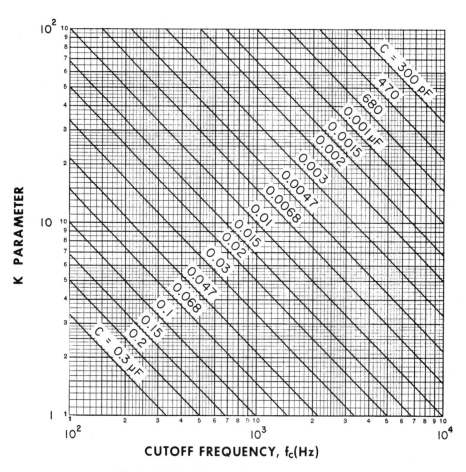

Fig. 2.12(*b*)  *K* parameter versus frequency.

21

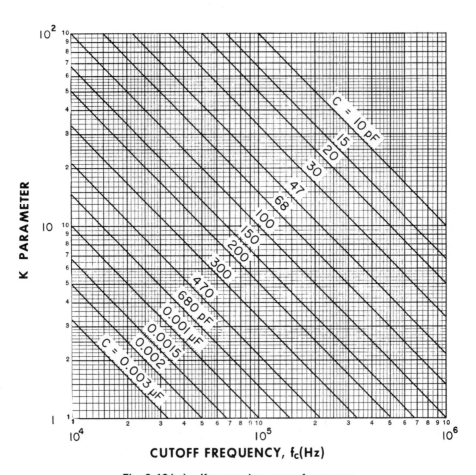

**CUTOFF FREQUENCY, $f_c$(Hz)**

**Fig. 2.12($c$)** *K* parameter versus frequency.

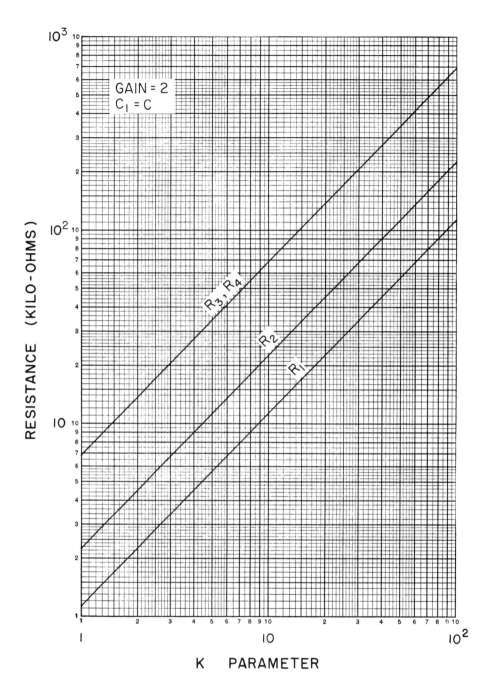

Fig. 2.13 Second-order VCVS low-pass Butterworth filter.

23

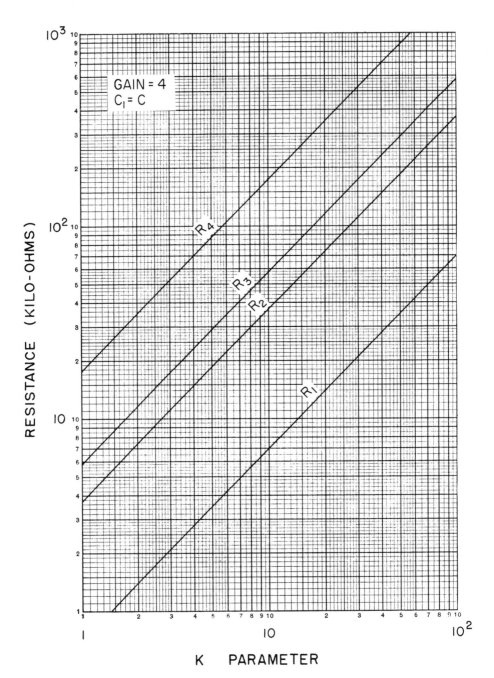

GAIN = 4
$C_I = C$

$R_4$
$R_3$
$R_2$
$R_1$

RESISTANCE (KILO-OHMS)

K   PARAMETER

Fig. 2.14   Second-order VCVS low-pass Butterworth filter.

24

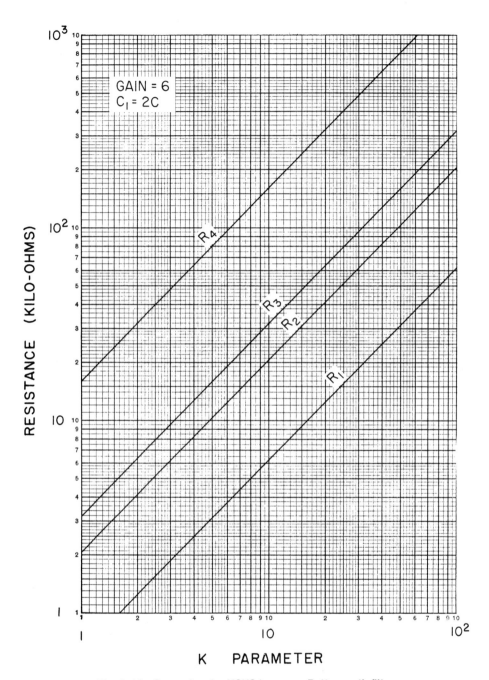

RESISTANCE (KILO-OHMS)

GAIN = 6
$C_1 = 2C$

$R_4$

$R_3$

$R_2$

$R_1$

K  PARAMETER

**Fig. 2.15  Second-order VCVS low-pass Butterworth filter.**

25

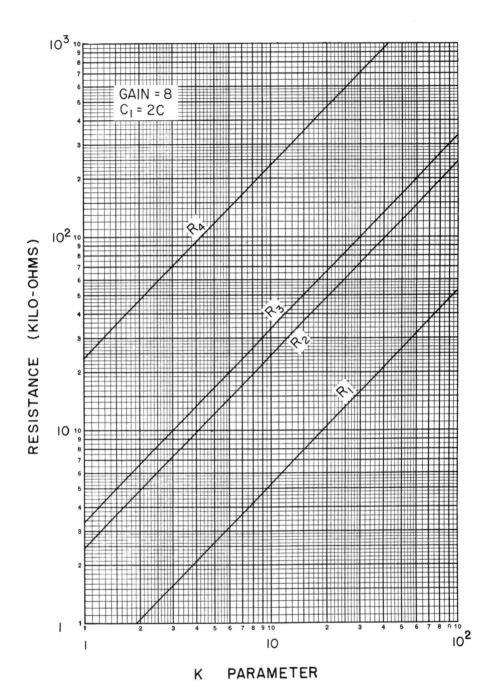

GAIN = 8
$C_1 = 2C$

RESISTANCE (KILO-OHMS)

$R_4$

$R_3$

$R_2$

$R_1$

K    PARAMETER

**Fig. 2.16   Second-order VCVS low-pass Butterworth filter.**

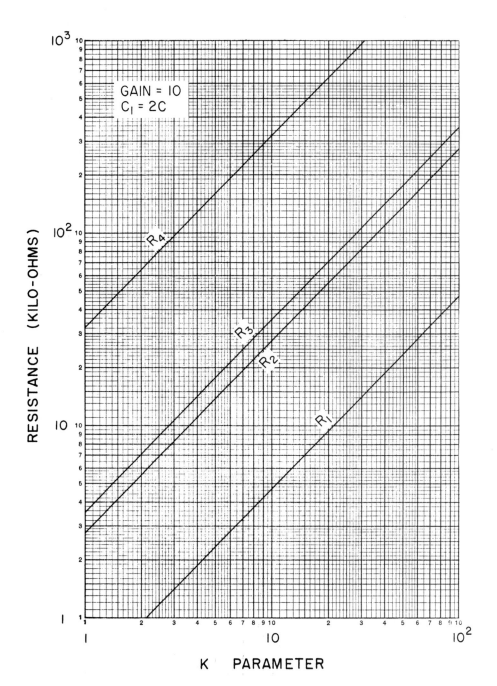

GAIN = 10
$C_1 = 2C$

RESISTANCE (KILO-OHMS)

$R_4$

$R_3$

$R_2$

$R_1$

K   PARAMETER

**Fig. 2.17   Second-order VCVS low-pass Butterworth filter.**

27

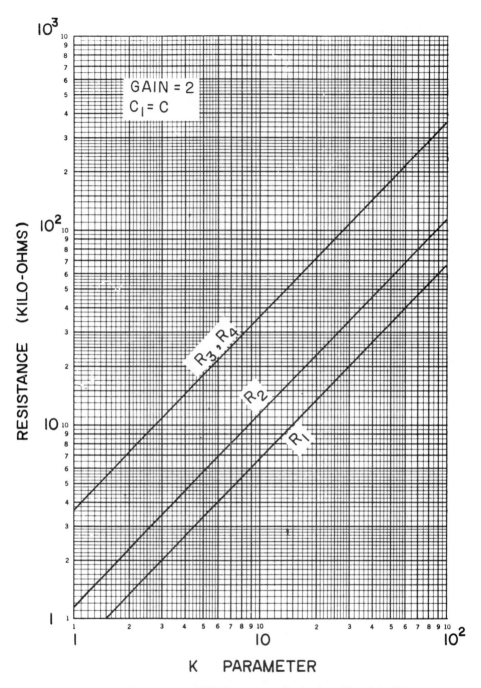

GAIN = 2
$C_1 = C$

$R_3$, $R_4$

$R_2$

$R_1$

RESISTANCE   (KILO-OHMS)

K   PARAMETER

Fig. 2.18   Second-order VCVS low-pass Chebyshev filter ( 1/10 dB).

28

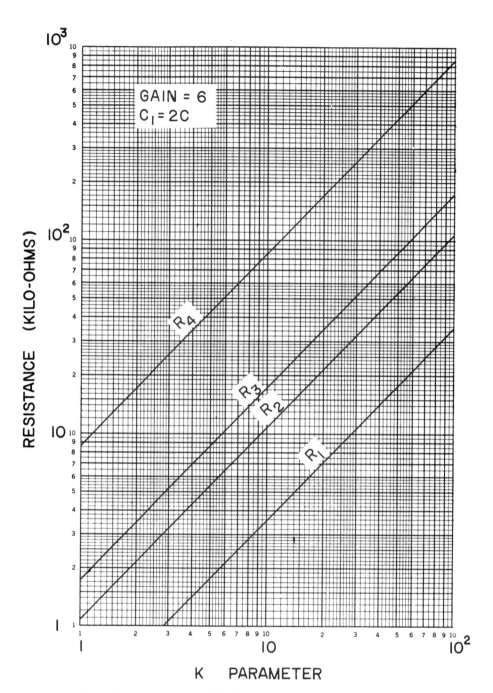

Fig. 2.19　Second-order VCVS low-pass Chebyshev filter ($\frac{1}{10}$ dB).

29

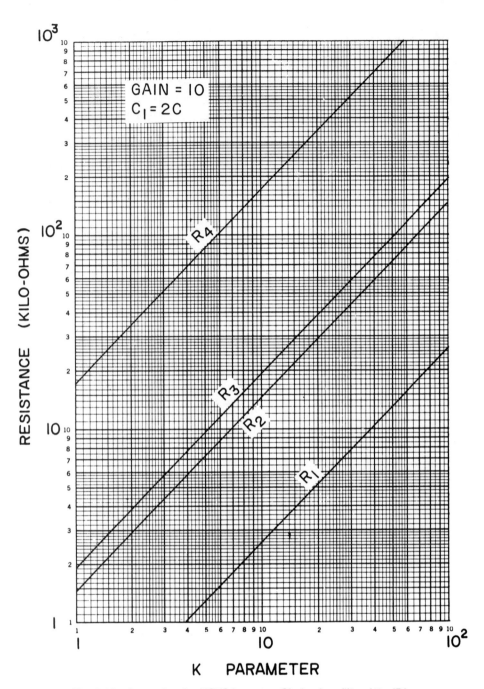

GAIN = 10
$C_1 = 2C$

RESISTANCE (KILO-OHMS)

$R_4$

$R_3$

$R_2$

$R_1$

K PARAMETER

Fig. 2.20 Second-order VCVS low-pass Chebyshev filter ($\frac{1}{10}$ dB).

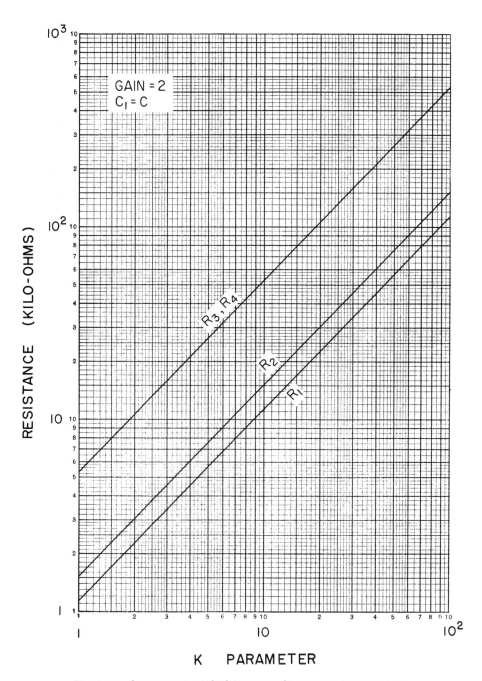

## K PARAMETER

Fig. 2.21  Second-order VCVS low-pass Chebyshev filter ( ½ dB).

31

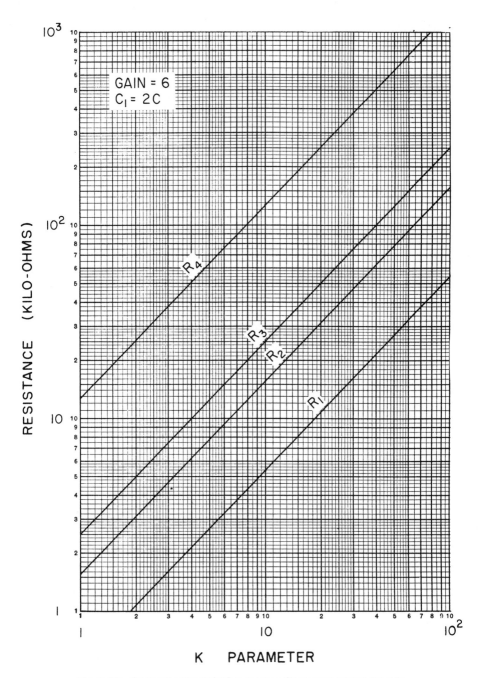

Fig. 2.22 Second-order VCVS low-pass Chebyshev filter ( ½ dB ).

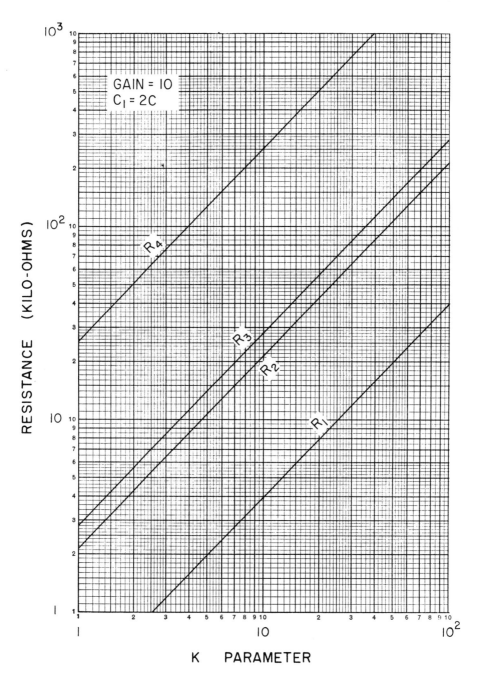

**RESISTANCE (KILO-OHMS)**

GAIN = 10
$C_I = 2C$

$R_4$

$R_3$

$R_2$

$R_1$

**K   PARAMETER**

Fig. 2.23  Second-order VCVS low-pass Chebyshev filter ( ½ dB ).

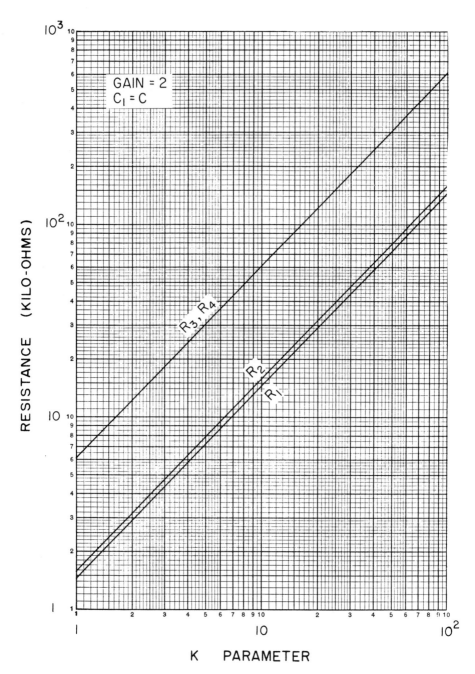

**Fig. 2.24  Second-order VCVS low-pass Chebyshev filter ( 1 dB).**

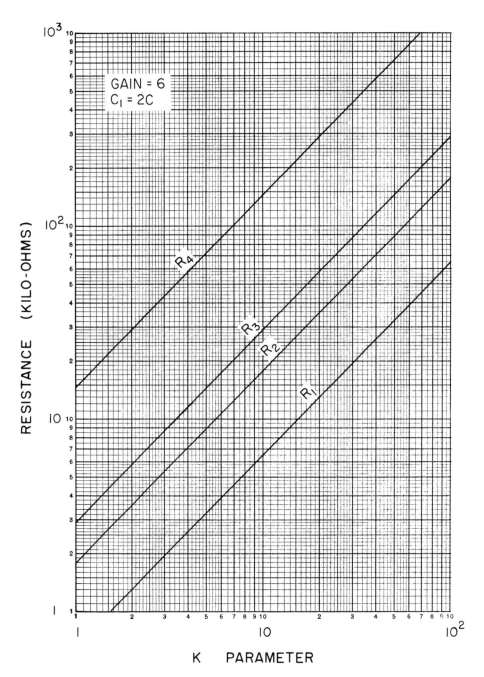

**Fig. 2.25 Second-order VCVS low-pass Chebyshev filter ( 1 dB).**

35

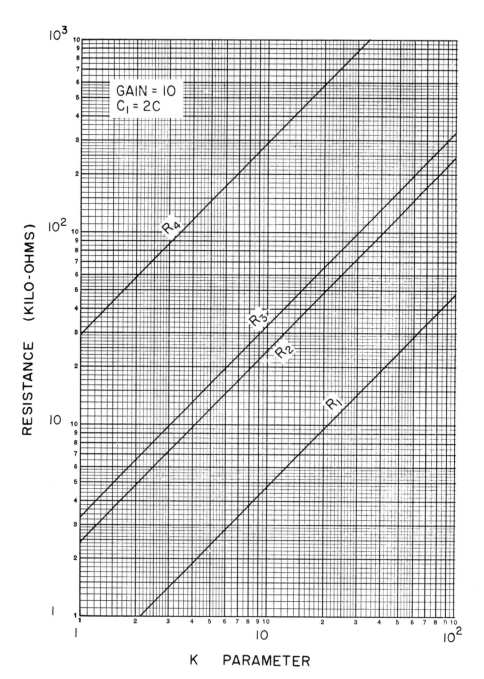

GAIN = 10
$C_1 = 2C$

RESISTANCE (KILO-OHMS)

K PARAMETER

Fig. 2.26 Second-order VCVS low-pass Chebyshev filter ( 1 dB).

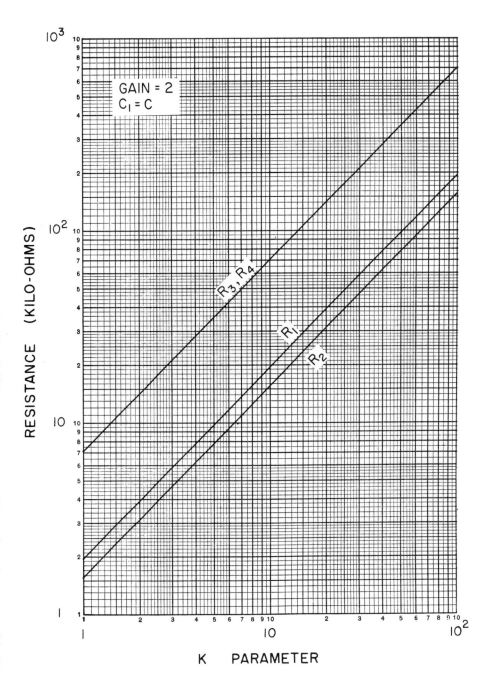

**RESISTANCE (KILO-OHMS)**

GAIN = 2
$C_I = C$

$R_3, R_4$

$R_1$

$R_2$

**K   PARAMETER**

Fig. 2.27   Second-order VCVS low-pass Chebyshev filter (2 dB).

37

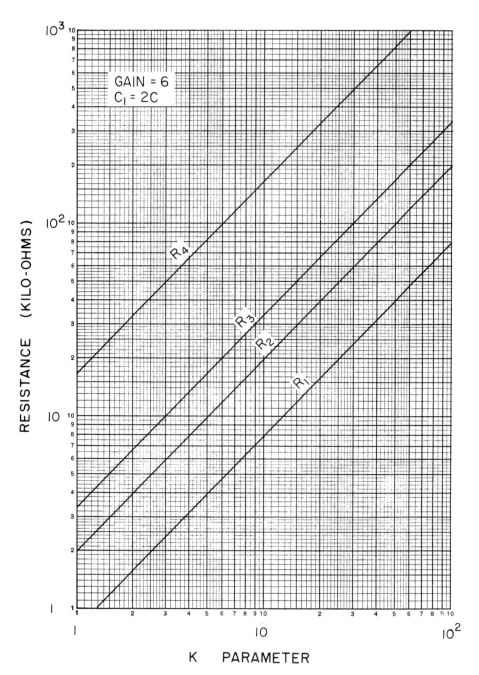

GAIN = 6
$C_1 = 2C$

Fig. 2.28  Second-order VCVS low-pass Chebyshev filter ( 2 dB).

RESISTANCE  (KILO-OHMS)

K    PARAMETER

$R_4$

$R_3$

$R_2$

$R_1$

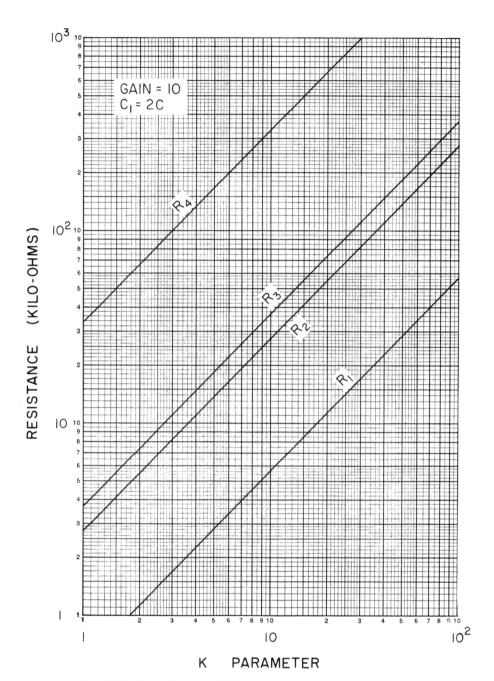

Fig. 2.29 Second-order VCVS low-pass Chebyshev filter (2 dB).

39

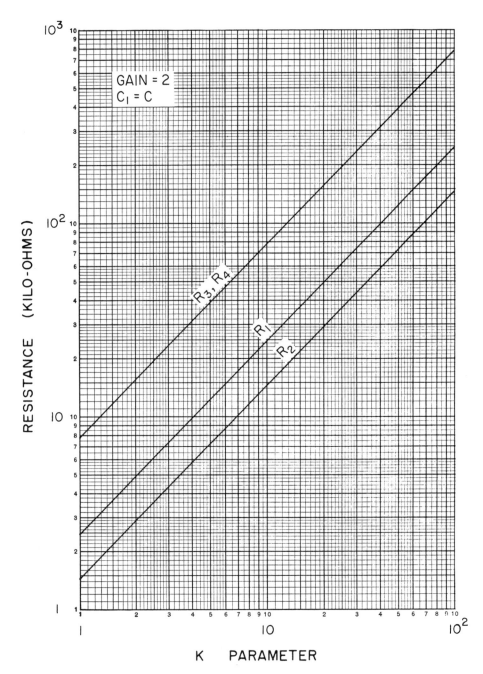

GAIN = 2
$C_1 = C$

RESISTANCE (KILO-OHMS)

K    PARAMETER

Fig. 2.30   Second-order VCVS low-pass Chebyshev filter (3 dB).

40

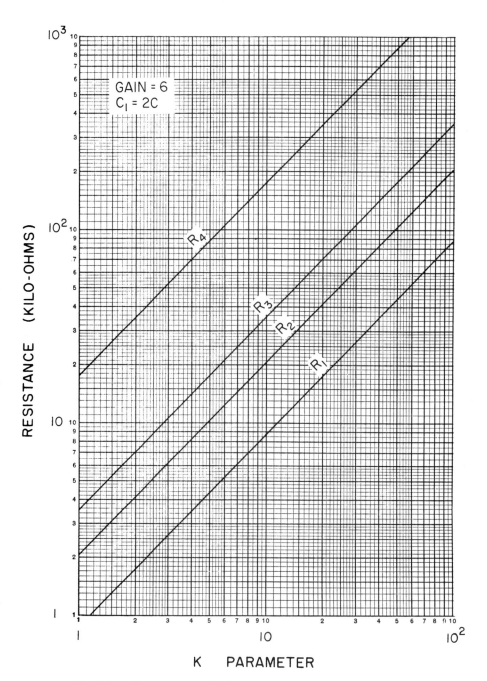

**Fig. 2.31   Second-order VCVS low-pass Chebyshev filter ( 3 dB).**

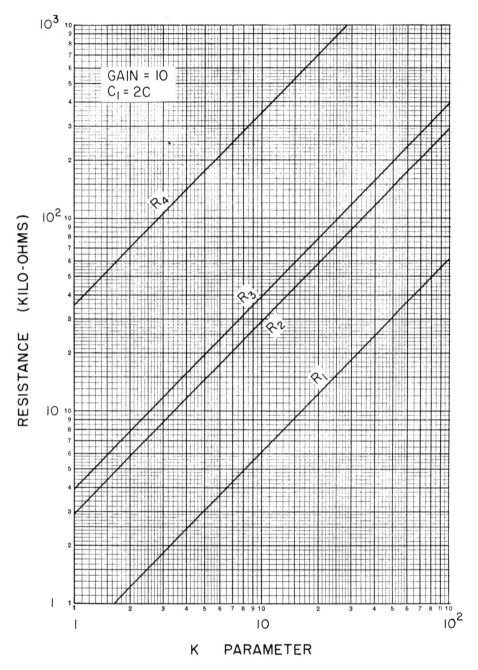

RESISTANCE (KILO-OHMS)

GAIN = 10
$C_I = 2C$

K    PARAMETER

Fig. 2.32  Second-order VCVS low-pass Chebyshev filter (3 dB).

42

## SUMMARY OF LOW-PASS SECOND-ORDER BIQUAD FILTER DESIGN PROCEDURE

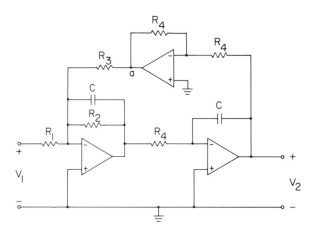

**General circuit.**

### Procedure

Given $f_c$ (Hz), gain, and filter type (Butterworth or Chebyshev), perform the following steps:

1. Select a value of capacitance $C$, determining a $K$ parameter from Fig. 2.12 $a$, $b$, or $c$, as described in the second-order VCVS low-pass case.
2. Using this value of $K$ and the given gain, find the resistance values of the circuit from the appropriate one of Fig. 2.33 for the Butterworth filter and Figs. 2.34 through 2.38 for the Chebyshev filter, depending, in the Chebyshev case, on the dB ripple desired.
3. Select standard resistance values which are as close as possible to those indicated on the graph and construct the circuit.

### Comments and Suggestions

The suggestions given in the second-order VCVS low-pass case apply except that there are three op-amps instead of one and there is no resistance ratio to be used in minimizing the dc offset of the op-amps. Also the dc return to ground requirement is satisfied by resistors $R_2$ and $R_3$. The gain is $R_3/R_1$ and an inverting gain (negative) of the same magnitude may be obtained by taking the output at node $a$. The filter response is readily adjusted by varying $R_1$, $R_2$, and $R_3$. Varying $R_1$ affects the gain, varying $R_2$ affects the passband response, and varying $R_3$ affects $f_c$.

The second-order biquad low-pass circuit is discussed in Sec. 2.1.

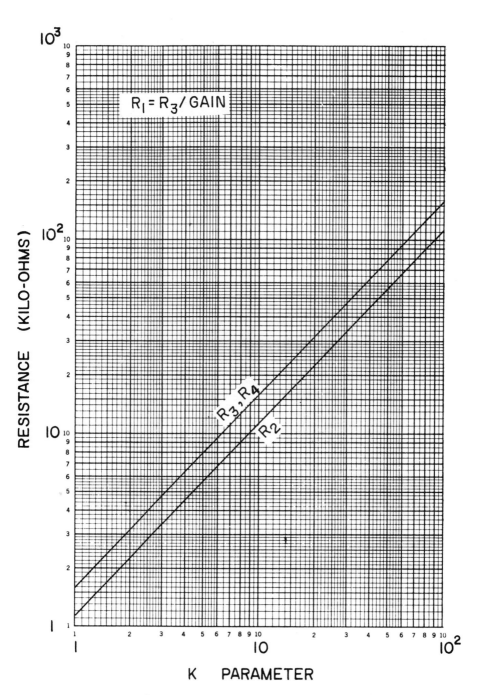

Fig. 2.33  Second-order biquad low-pass Butterworth filter.

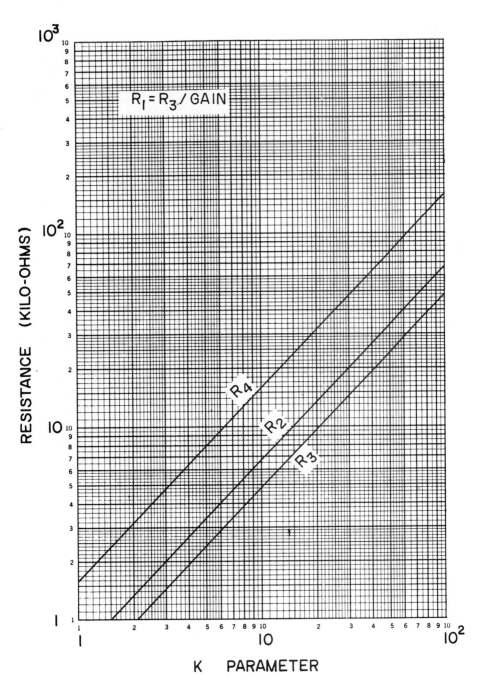

Fig. 2.34  Second-order biquad low-pass Chebyshev filter ( $\frac{1}{10}$ dB).

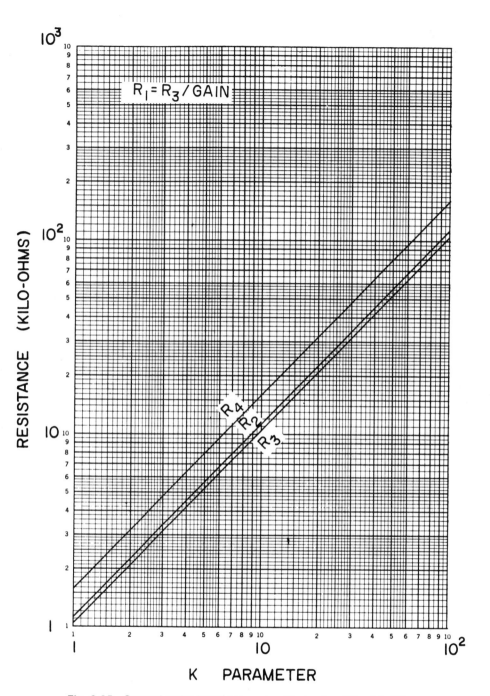

Fig. 2.35   Second-order biquad low-pass Chebyshev filter ( ½ dB).

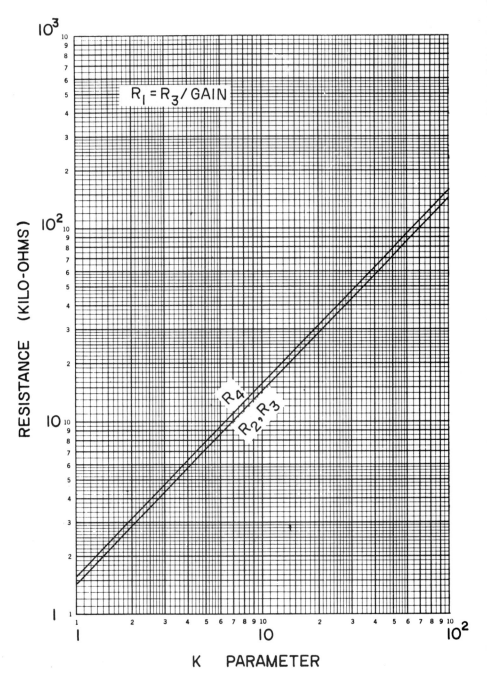

Fig. 2.36 Second-order biquad low-pass Chebyshev filter ( 1 dB).

48

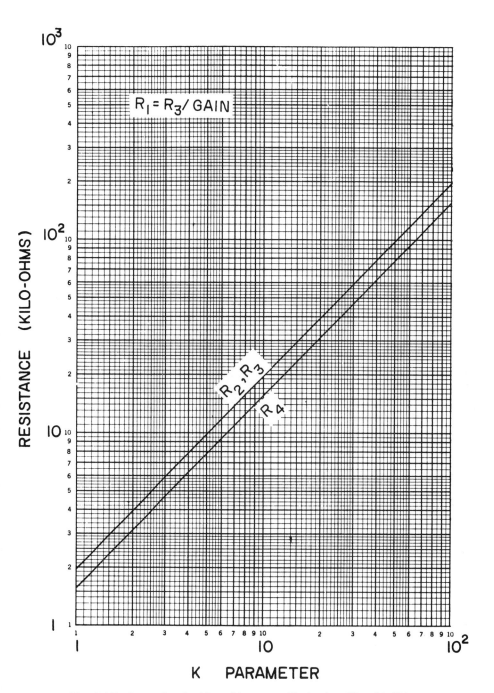

Fig. 2.37  Second-order biquad low-pass Chebyshev filter ( 2 dB).

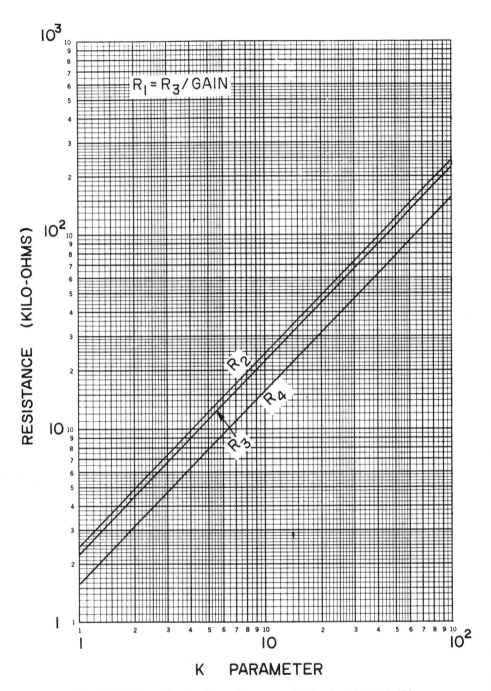

Fig. 2.38 Second-order biquad low-pass Chebyshev filter (3 dB).

## SUMMARY OF LOW-PASS FOURTH-ORDER VCVS
## FILTER DESIGN PROCEDURE

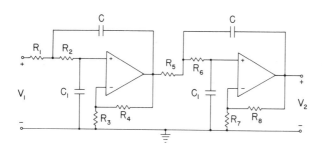

**General Circuit.**

## Procedure

Given $f_c$ (Hz), gain, and filter type (Butterworth or Chebyshev), perform the following steps:

1. Select a value of capacitance $C$, determining a $K$ parameter from Fig. 2.12$a$, $b$, or $c$, as described in the second-order VCVS low-pass case.
2. Using this value of $K$ find the remaining element values of the circuit from the appropriate one of Figs. 2.39 through 2.41 for the Butterworth filter, and Figs. 2.42 through 2.56 for the Chebyshev filter, depending on the gain and, in the Chebyshev case, the dB ripple desired.
3. Select standard element values which are as close as possible to those indicated on the graph and construct the circuit.

## Comments and Suggestions

The suggestions given in the second-order VCVS low-pass case apply except that the open-loop gain of the op-amps should be at least 50 times the square root of the filter gain. The remarks in the second-order case for $R_3$ and $R_4$ apply also to $R_7$ and $R_8$.
   A specific example is given in Sec. 2.6.

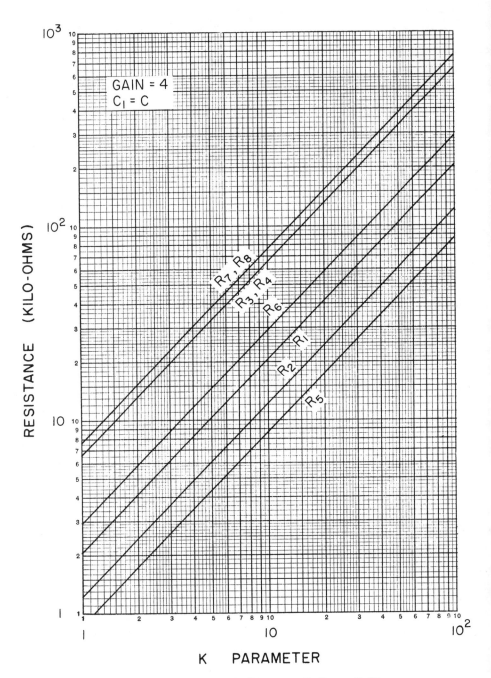

Fig. 2.39 Fourth-order VCVS low-pass Butterworth filter.

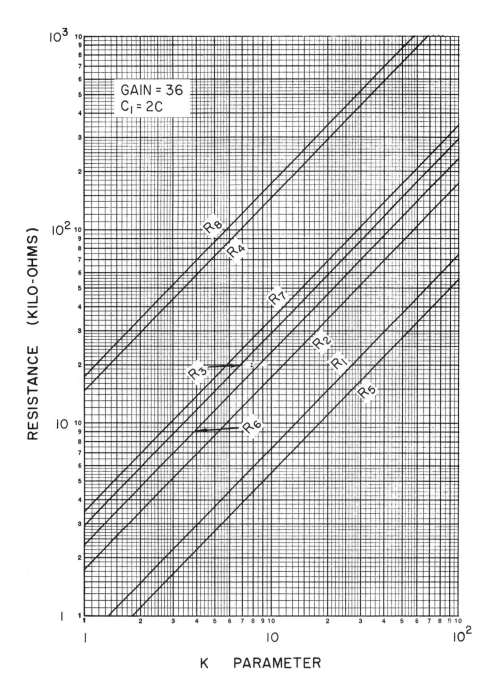

GAIN = 36
$C_I = 2C$

$10^3$

$10^2$

$10$

$1$

RESISTANCE (KILO-OHMS)

$R_8$
$R_4$
$R_7$
$R_2$
$R_3$
$R_1$
$R_5$
$R_6$

$1$

$10$

$10^2$

K   PARAMETER

Fig. 2.40   Fourth-order VCVS low-pass Butterworth filter.

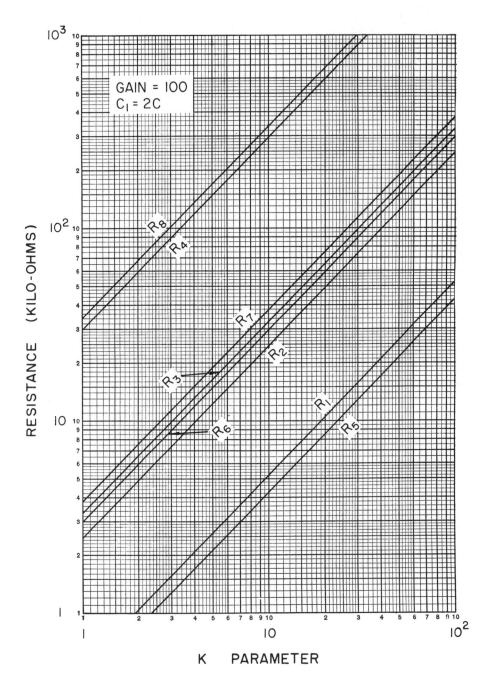

GAIN = 100
$C_I = 2C$

**RESISTANCE (KILO-OHMS)**

$R_8$
$R_4$
$R_7$
$R_2$
$R_3$
$R_1$
$R_5$
$R_6$

**K PARAMETER**

Fig. 2.41 Fourth-order VCVS low-pass Butterworth filter.

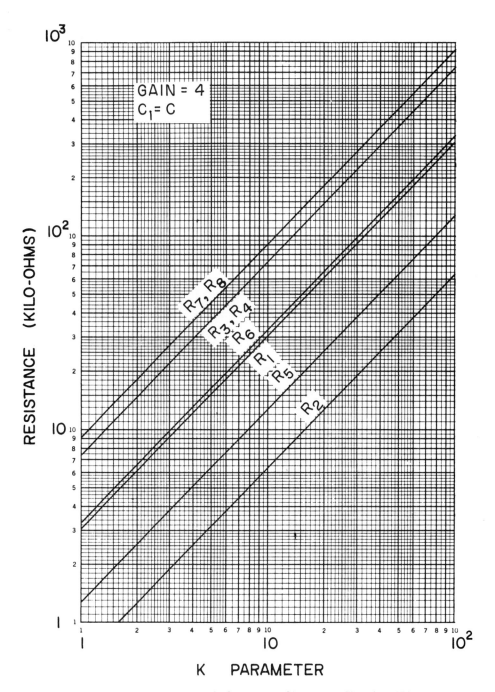

**Fig. 2.42   Fourth-order VCVS low-pass Chebyshev filter ( ⅟₁₀ dB ).**

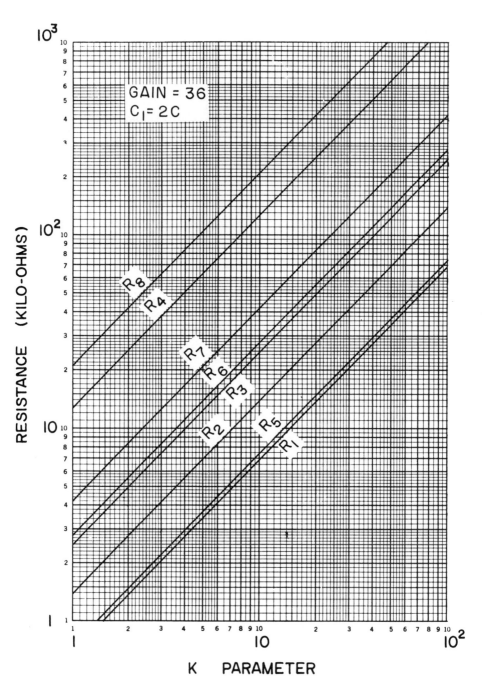

Fig. 2.43 Fourth-order VCVS low-pass Chebyshev filter (1/10 dB).

56

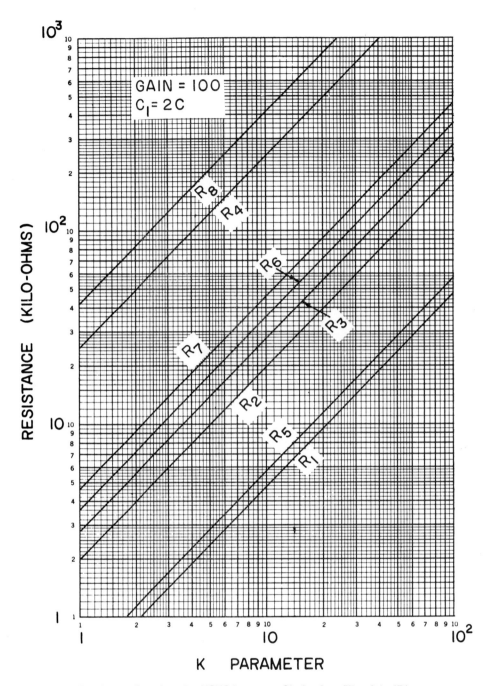

Fig. 2.44 Fourth-order VCVS low-pass Chebyshev filter (1/10 dB).

57

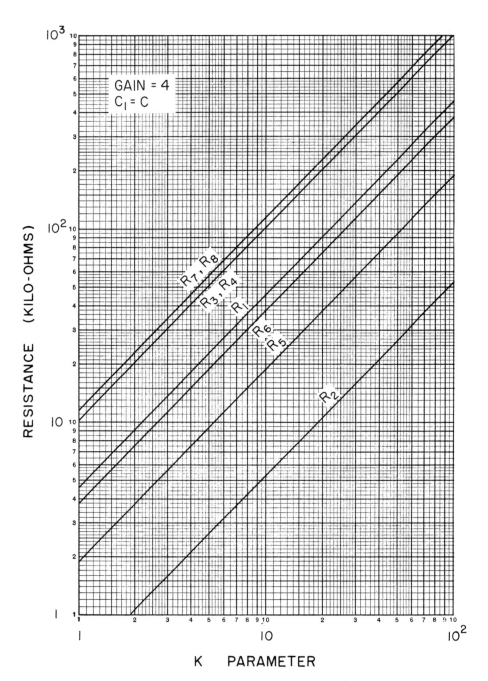

Fig. 2.45   Fourth-order VCVS low-pass Chebyshev filter ( ½ dB).

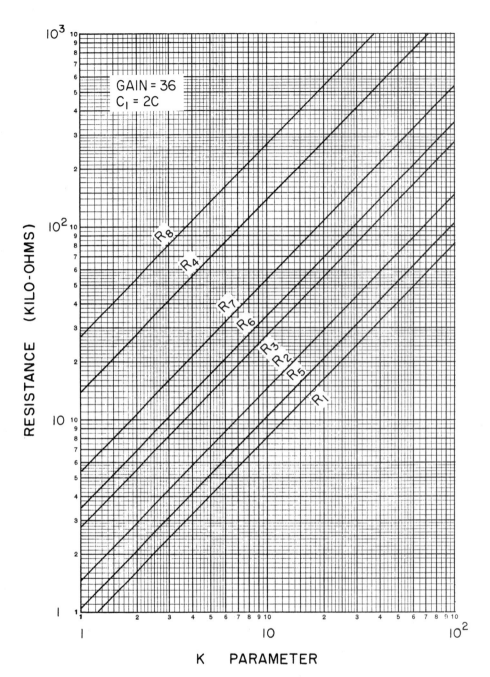

**Fig. 2.46   Fourth-order VCVS low-pass Chebyshev filter ( ½ dB).**

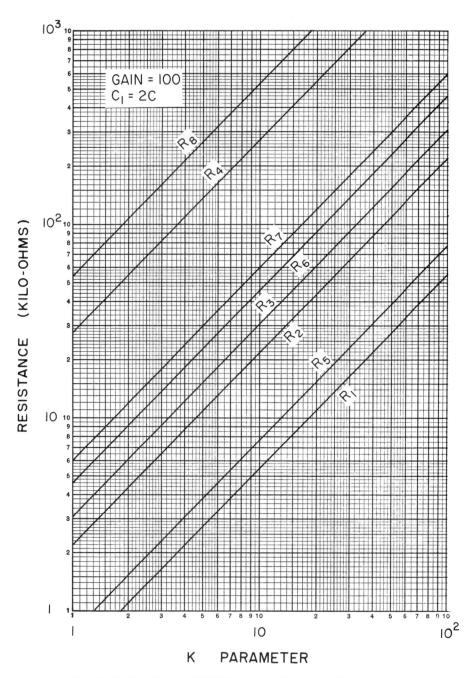

Fig. 2.47   Fourth-order VCVS low-pass Chebyshev filter ( ½ dB).

60

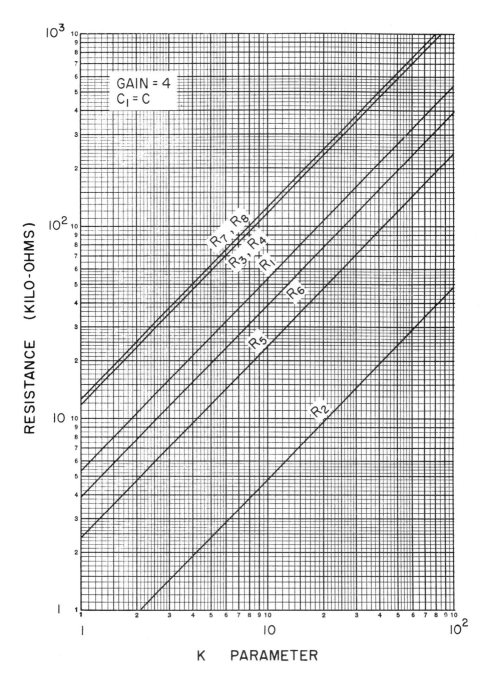

GAIN = 4
$C_I = C$

RESISTANCE (KILO-OHMS)

$R_7, R_8$
$R_3, R_4$
$R_1$
$R_6$
$R_5$
$R_2$

K   PARAMETER

Fig. 2.48   Fourth-order VCVS low-pass Chebyshev filter (1 dB).

61

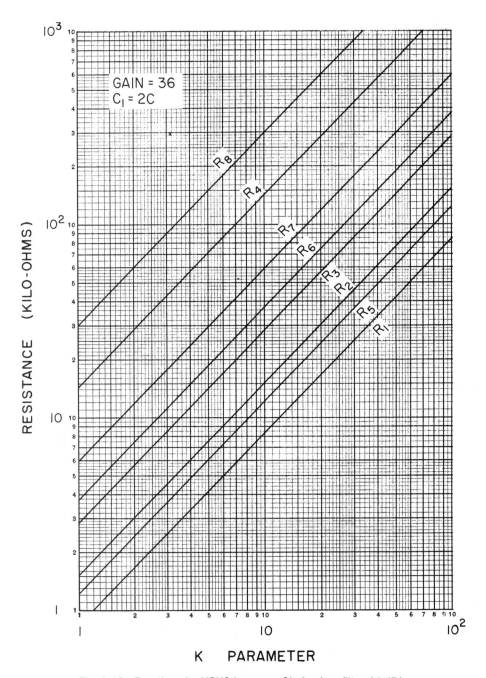

Fig. 2.49   Fourth-order VCVS low-pass Chebyshev filter ( 1 dB).

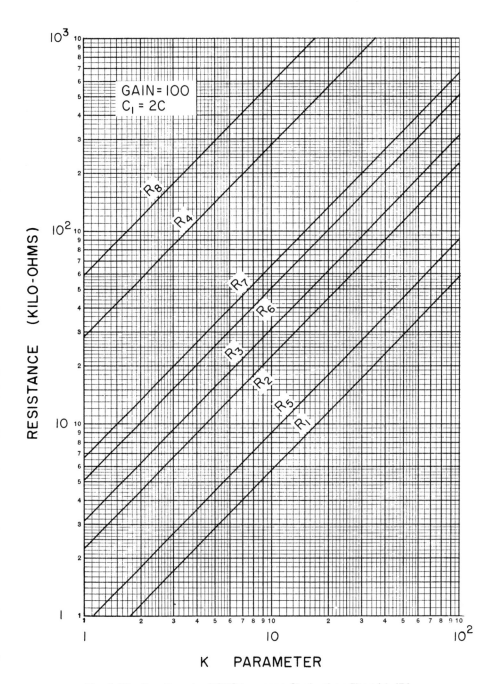

RESISTANCE (KILO-OHMS)

GAIN = 100
$C_1 = 2C$

K   PARAMETER

**Fig. 2.50   Fourth-order VCVS low-pass Chebyshev filter ( 1 dB ).**

63

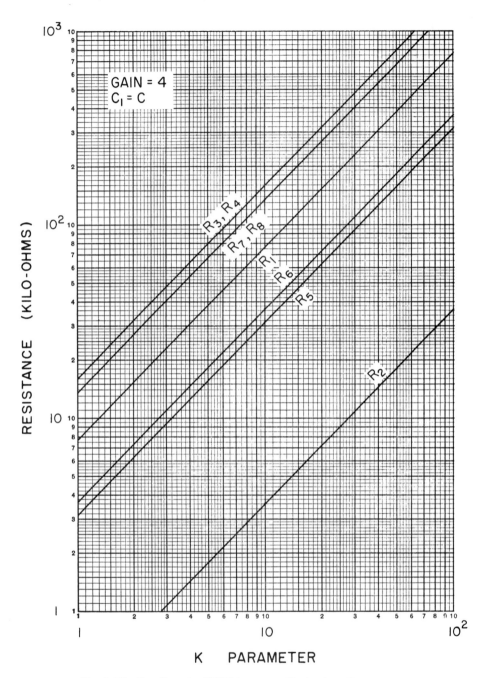

Fig. 2.51 Fourth-order VCVS low-pass Chebyshev filter (2 dB).

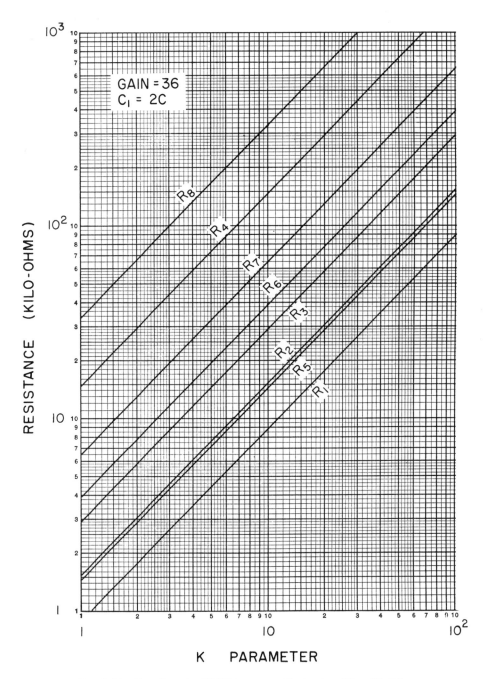

GAIN = 36
$C_1 = 2C$

RESISTANCE (KILO-OHMS)

$R_8$ $R_4$ $R_7$ $R_6$ $R_3$ $R_2$ $R_5$ $R_1$

K    PARAMETER

Fig. 2.52   Fourth-order VCVS low-pass Chebyshev filter ( 2 dB).

65

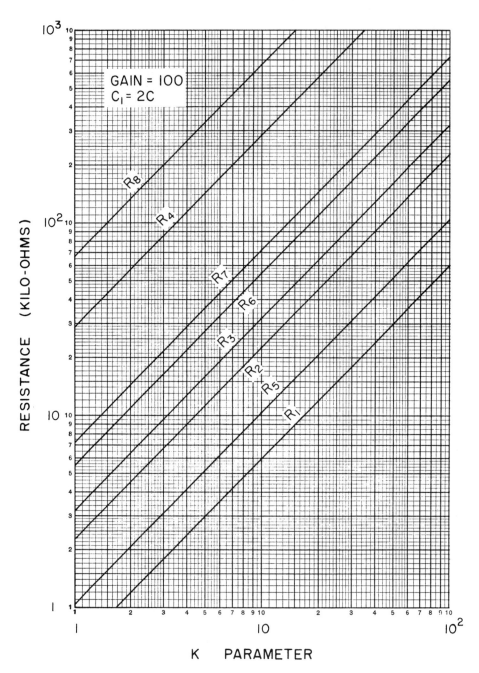

RESISTANCE (KILO-OHMS)

GAIN = 100
$C_1 = 2C$

$R_8$
$R_4$
$R_7$
$R_6$
$R_3$
$R_2$
$R_5$
$R_1$

K    PARAMETER

Fig. 2.53   Fourth-order VCVS low-pass Chebyshev filter (2 dB).

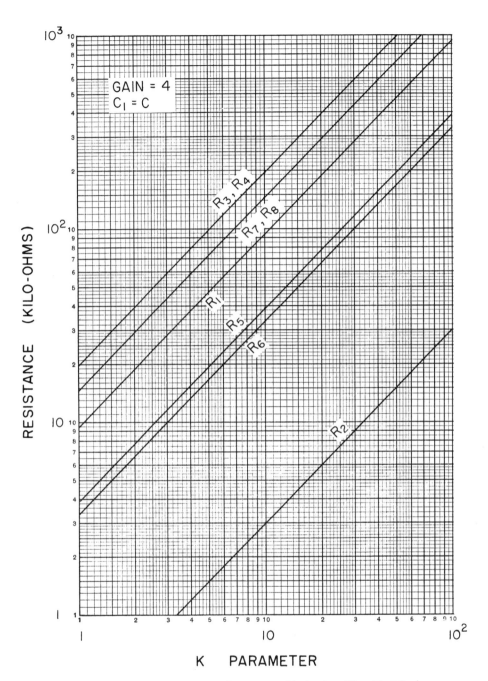

Fig. 2.54  Fourth-order VCVS low-pass Chebyshev filter (3 dB).

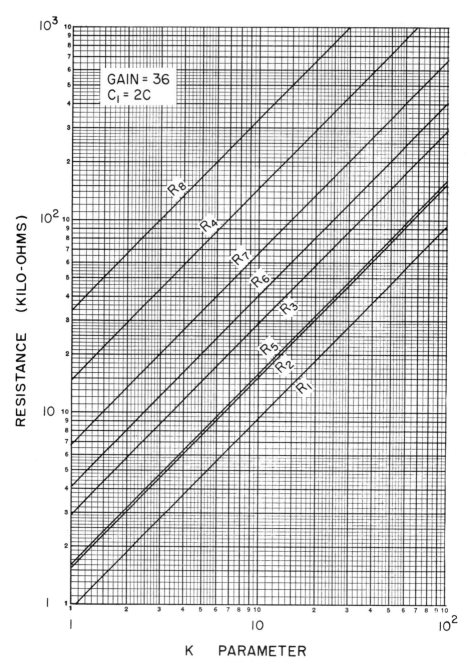

Fig. 2.55  Fourth-order VCVS low-pass Chebyshev filter ( 3 dB ).

68

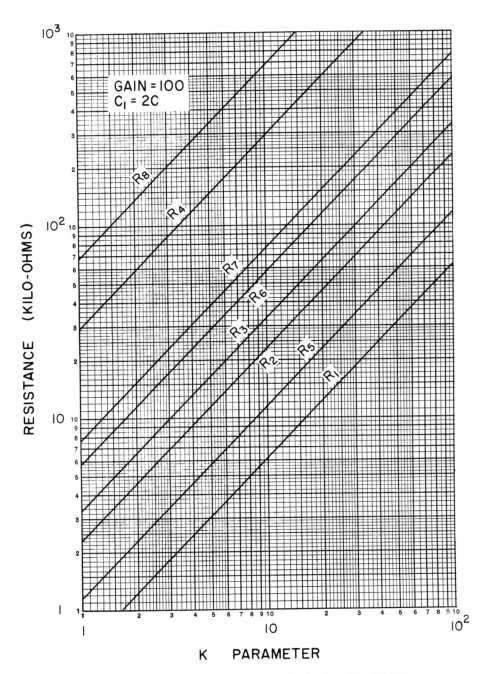

Fig. 2.56 Fourth-order VCVS low-pass Chebyshev filter (3 dB).

## SUMMARY OF LOW-PASS FOURTH-ORDER BIQUAD FILTER DESIGN PROCEDURE

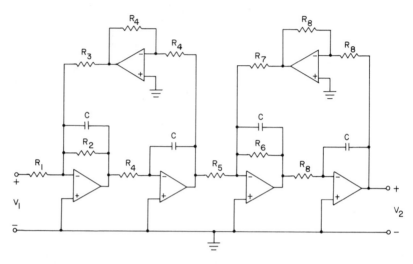

**General circuit.**

### Procedure

Given $f_c$ (Hz), gain, and filter type (Butterworth or Chebyshev), perform the following steps:

1. Select a value of $C$, determining a $K$ parameter from Fig. 2.12*a*, *b*, or *c*, as described in the second-order VCVS low-pass case.
2. Using this value of $K$ and the given gain, find the resistance values of the circuit from the appropriate one of Fig. 2.57 for the Butterworth filter and Figs. 2.58 through 2.62 for the Chebyshev filter, depending, in the Chebyshev case, on the dB ripple desired.
3. Select standard resistance values which are as close as possible to those indicated on the graph and construct the circuit.

### Comments and Suggestions

The suggestions given in the second-order VCVS low-pass case apply except that there are six op-amps instead of one and there is no resistance ratio to be used in minimizing the dc offset of the op-amps. Also the dc return to ground requirement is satisfied by resistors $R_2$, $R_3$, $R_6$, and $R_7$.

The graphs are drawn for each stage to have the square root of the gain ($\sqrt{\text{GAIN}}$). The gain may be distributed differently by setting the gain of the first stage at $G_1$ ($R_1 = R_3/G_1$) and the gain of the second stage at $G_2$ ($R_5 = R_7/G_2$), so long as the product of $G_1$ and $G_2$ is the filter gain (GAIN).

The filter response may be adjusted by varying $R_1$ and $R_5$ to affect the gains, varying $R_2$ and $R_6$ to affect the passband response, and varying $R_3$ and $R_7$ to affect $f_c$.

The fourth-order biquad low-pass circuit is discussed in Sec. 2.6.

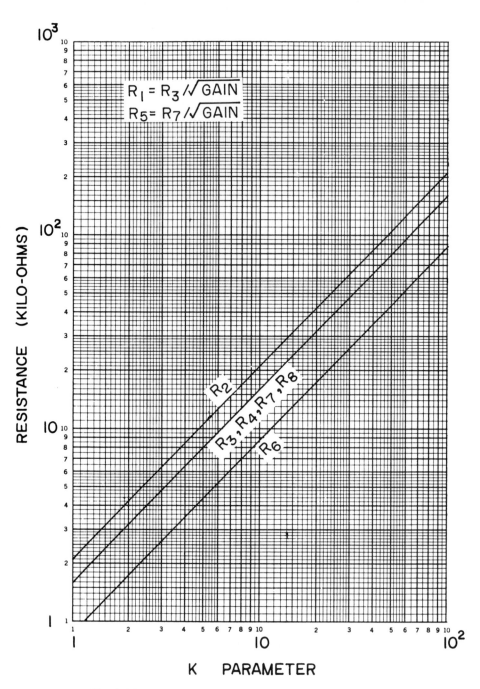

Fig. 2.57 Fourth-order biquad low-pass Butterworth filter.

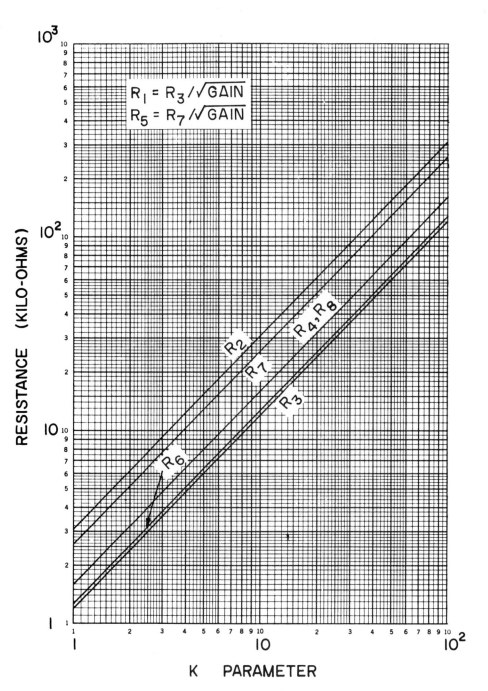

$$R_1 = R_3 / \sqrt{GAIN}$$
$$R_5 = R_7 / \sqrt{GAIN}$$

RESISTANCE (KILO-OHMS)

K   PARAMETER

Fig. 2.58   Fourth-order biquad low-pass Chebyshev filter ($\frac{1}{10}$ dB).

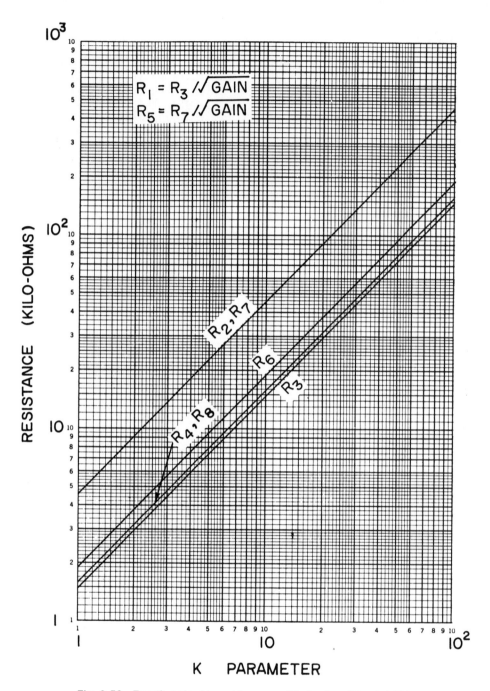

Fig. 2.59 Fourth-order biquad low-pass Chebyshev filter ( ½ dB).

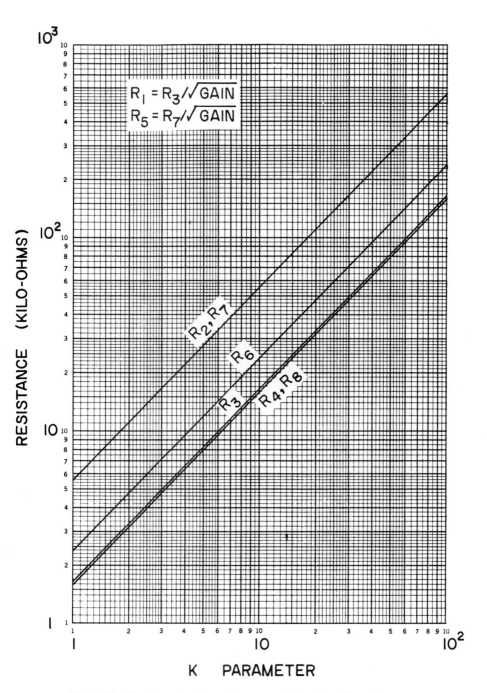

Fig. 2.60　Fourth-order biquad low-pass Chebyshev filter ( 1 dB).

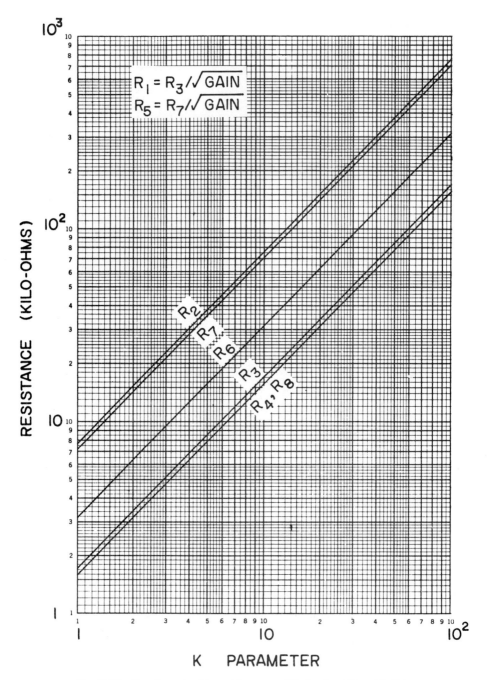

$$R_1 = R_3/\sqrt{GAIN}$$
$$R_5 = R_7/\sqrt{GAIN}$$

RESISTANCE (KILO-OHMS)

K PARAMETER

Fig. 2.61 Fourth-order biquad low-pass Chebyshev filter (2 dB).

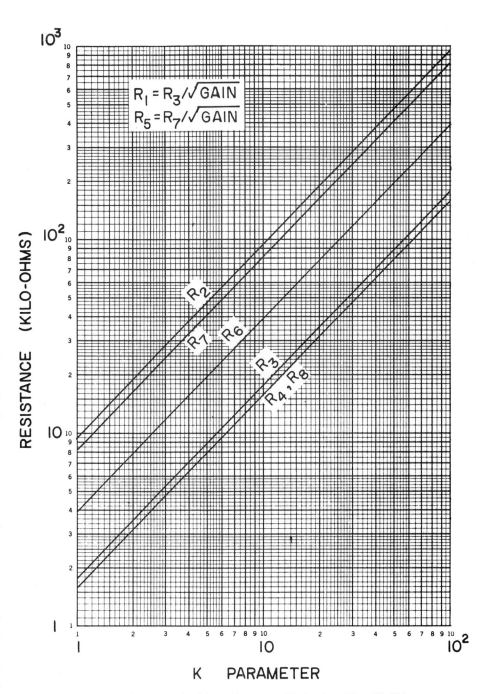

Fig. 2.62   Fourth-order biquad low-pass Chebyshev filter (3 dB).

# 3

## HIGH-PASS FILTERS

### 3.1 General Circuit and Equations

A high-pass filter passes high frequencies and attenuates low frequencies, as shown by the amplitude response of Fig. 3.1. The response represented by the broken line is ideal, and the response represented by the solid line is a realizable approximation to the ideal.

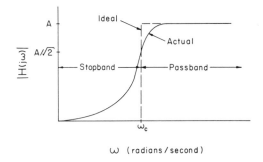

Fig. 3.1 A high-pass filter response.

The second-order approximation to the ideal is achieved, for appropriate values of $a$ and $b$, by the transfer function [21]

$$H(s) = \frac{V_2(s)}{V_1(s)} = \frac{Ks^2}{s^2 + as + b} \tag{3.1}$$

An $n$th-order transfer function has numerator $Ks^n$ and an $n$th degree polynomial denominator. For example, a fourth-order function is the product of two second-order functions such as Eq. (3.1).

The cutoff frequency is $\omega_c$ in radians/sec or $f_c = \omega_c/2\pi$ Hz. The gain of the filter is the value of $H(s)$ as $s$ becomes infinite, which is evidently $K$. The high-pass Butterworth filter amplitude response is monotonic in both the stop and passband and the high-pass Chebyshev filter amplitude response has ripples in its passband. Both are obtained from the low-pass amplitude functions by replacing $\omega$ by $1/\omega$ [24]. As in the low-pass case, the Chebyshev exhibits better cutoff characteristics but the Butterworth has a flatter passband characteristic.

In the case of the Butterworth filter, we use $f_c$ as the conventional cutoff point, but in the Chebyshev high-pass filter we use $f_c$ as the beginning of the ripple channel. The relationship between cutoff in the conventional sense, and $f_c$ in this case is given in Secs. 3.2 and 3.3 for the second- and fourth-order filters respectively.

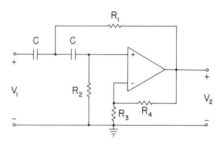

**Fig. 3.2   A high-pass filter.**

A network which gives, for suitable values of the resistances and capacitances, either a second-order Butterworth or a second-order Chebyshev high-pass filter is the Sallen and Key VCVS circuit [17] shown in Fig. 3.2. Analysis of this circuit shows that it realizes Eq. (3.1) for the values

$$K = \mu = 1 + \frac{R_4}{R_3}$$

$$a = \frac{1}{R_1 C}(1 - \mu) + \frac{2}{R_2 C} \tag{3.2}$$

$$b = \frac{1}{R_1 R_2 C^2}$$

Figure 3.2 is the circuit we shall use for second-order high-pass filters, and for the fourth-order case we shall cascade two such networks.

## 3.2  Second-Order High-Pass Filters

We shall use our graphs as described in the low-pass Butterworth development of Sec. 2.3 to obtain practical Butterworth and Chebyshev second-order high-pass filters. The design procedure is described in detail in the summary at the end of the chapter.

In the Butterworth case, $f_c$ is the conventional cutoff point, but in the Chebyshev case, $f_c$ is the beginning of the ripple channel. In the second-order Chebyshev high-pass filter, the cutoff point is 0.515 $f_c$ for the ¹⁄₁₀ dB filter, 0.719 $f_c$ for the ½ dB, 0.826 $f_c$ for the 1 dB, 0.935 $f_c$ for the 2 dB, and $f_c$ for the 3 dB.

As an example, suppose we want a 1 dB Chebyshev filter with $f_c$ = 100,000 Hz, gain = 2, and C = 200 pF. Then from Fig. 3.6c, we have the K parameter = 5, and for a gain of 2 we have from Fig. 3.19, $R_1 = R_2 =$ 8.3 kΩ and $R_3 = R_4 =$ 16.5 kΩ. Using the standard values of resistance, 7.5 and 16 kΩ, with a μA709 op-amp, compensated with an 11 pF capacitor and a series combination of a 200 pF capacitor and a 1.5 kΩ resistor, we obtain the response shown in Fig. 3.3. The actual results are $f_c$ = 98,400 Hz, gain = 2, and the ripple width is 1 dB. The response is plotted on a scale of 50,000 Hz/division.

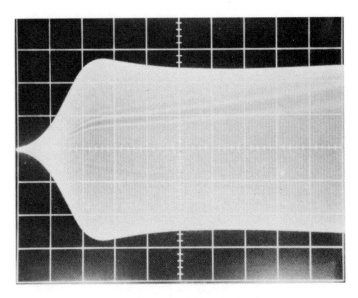

**Fig. 3.3  A second-order high-pass Chebyshev response.**

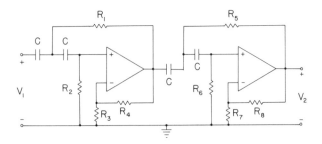

**Fig. 3.4    A fourth-order high-pass filter.**

## 3.3    Fourth-Order High-Pass Filters

To obtain either fourth-order Butterworth or Chebyshev high-pass filters, we use the circuit of Fig. 3.4, which is a cascading of two circuits of the type shown in Fig. 3.2. For a given $f_c$ and $C$, we may obtain a practical circuit as described in detail in the summary at the end of the chapter.

As in the second-order case, for the Butterworth filter $f_c$ is the cutoff point, but for the Chebyshev filter, $f_c$ is the beginning of the ripple channel. For the fourth-order high-pass Chebyshev filter the cutoff point is $0.824\,f_c$

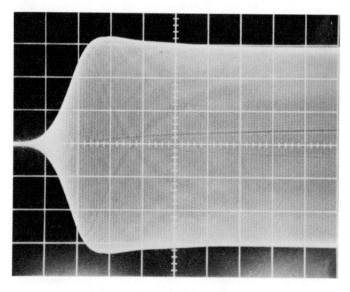

**Fig. 3.5    A fourth-order high-pass Butterworth response.**

for the $\frac{1}{10}$ dB case, 0.917 $f_c$ for the $\frac{1}{2}$ dB case, 0.952 $f_c$ for the 1 dB case, 0.980 $f_c$ for the 2 dB case, and $f_c$ for the 3 dB case.

As an example, we obtain a fourth-order Butterworth high-pass filter with $f_c$ = 10,000 Hz, $C$ = 200 pF, and a gain of 4. From Fig. 3.6c, we have the $K$ parameter = 50, and from Fig. 3.28, we have $R_1$ = 73 k$\Omega$, $R_2$ = 87 k$\Omega$, $R_3$ = $R_4$ = 174 k$\Omega$, $R_5$ = 104 k$\Omega$, $R_6$ = 62 k$\Omega$, and $R_7$ = $R_8$ = 123 k$\Omega$. Using standard resistances of 72, 87, 180, 100, 62, and 120 k$\Omega$, we obtain a filter with response shown in Fig. 3.5. Two $\mu$A709 op-amps were used in the two stages, both compensated as in the example of Sec. 3.2. The scale in the picture starts at 0 Hz and each division represents 5000 Hz. The actual results were $f_c$ = 9595 Hz, with a gain of 4.5.

A summary of the techniques for constructing high-pass filters is given following this section. The second-order filters, together with their graphs, are presented first, followed by the fourth-order filters and their graphs.

## SUMMARY OF HIGH-PASS SECOND-ORDER FILTER DESIGN PROCEDURE

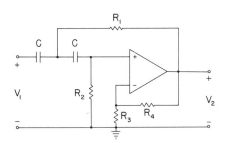

**General circuit.**

## Procedure

Given $f_c$ (Hz), gain, and filter type (Butterworth or Chebyshev), perform the following steps:

1. Select a value of capacitance $C$, determining a $K$ parameter from Fig. 3.6a if $f_c$ is between 1 and $10^2$ = 100, from Fig. 3.6b if $f_c$ is between 100 and $10^4$ = 10,000, and from Fig. 3.6c if $f_c$ is between 10,000 and $10^6$ = 1,000,000 Hz.

2. Using this value of $K$, find the resistance values of the circuit from the appropriate one of Figs. 3.7 through 3.12 for the Butterworth filter, and Figs. 3.13 through 3.27 for the

Chebyshev filter, depending on the gain and, in the Chebyshev case, the dB ripple desired.

3. Select standard resistance values which are as close as possible to those indicated on the graph and construct the circuit.

## Comments and Suggestions

These are exactly like those of the low-pass second-order case, except that the dc return to ground is already satisfied by the resistor $R_2$.

A specific example is given in Sec. 3.2.

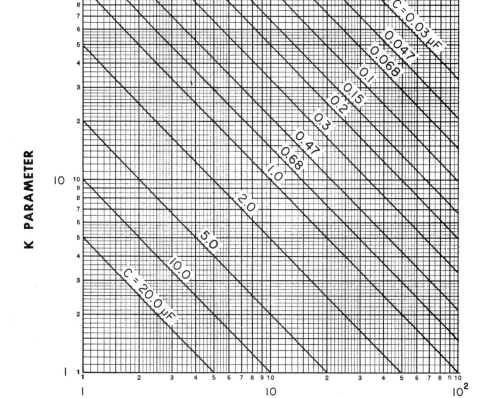

**K PARAMETER**

**CUTOFF FREQUENCY, f<sub>c</sub>(Hz)**

Fig. 3.6 ( *a* )  *K* parameter versus frequency.

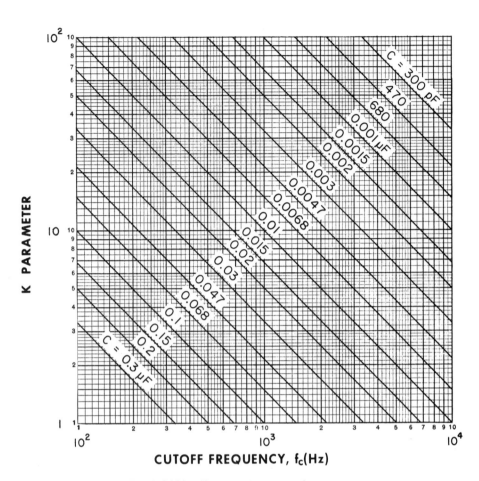

**K PARAMETER** (vertical axis label)

**CUTOFF FREQUENCY, $f_c$(Hz)**

Fig. 3.6(*b*)   *K* parameter versus frequency.

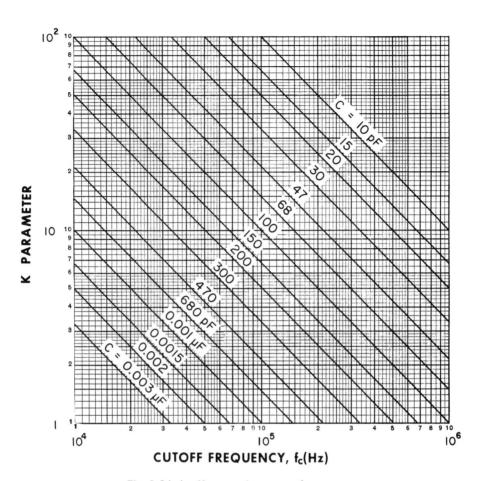

Fig. 3.6 ( c )   *K* parameter versus frequency.

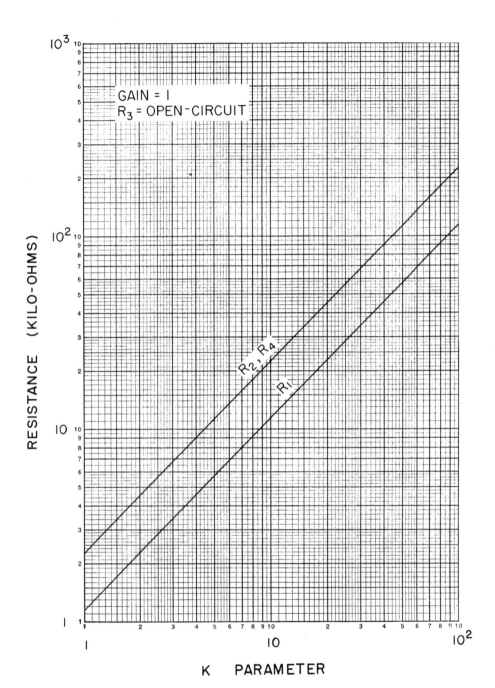

RESISTANCE (KILO-OHMS)

GAIN = 1
$R_3$ = OPEN-CIRCUIT

$R_2, R_4$

$R_1$

K    PARAMETER

Fig. 3.7    Second-order high-pass Butterworth filter.

88

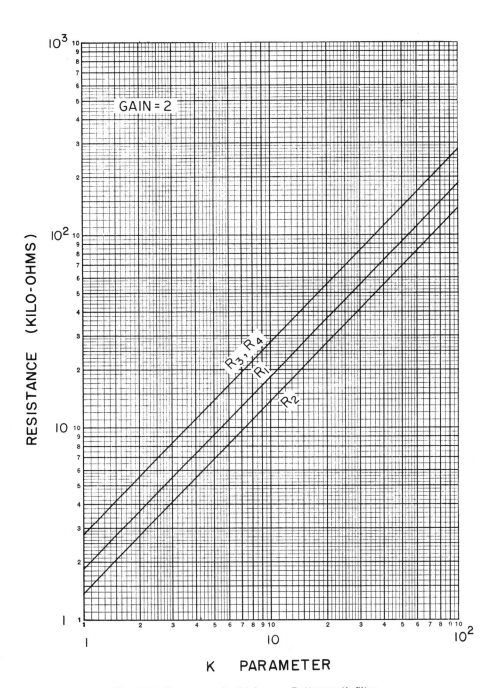

**Fig. 3.8  Second-order high-pass Butterworth filter.**

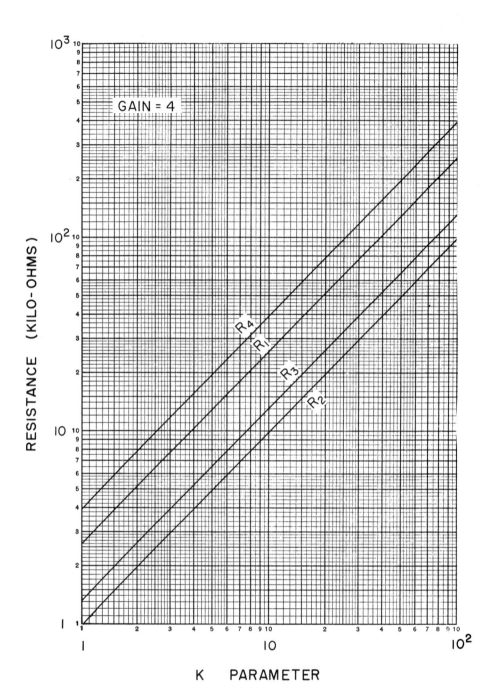

GAIN = 4

RESISTANCE (KILO-OHMS)

K    PARAMETER

**Fig. 3.9   Second-order high-pass Butterworth filter.**

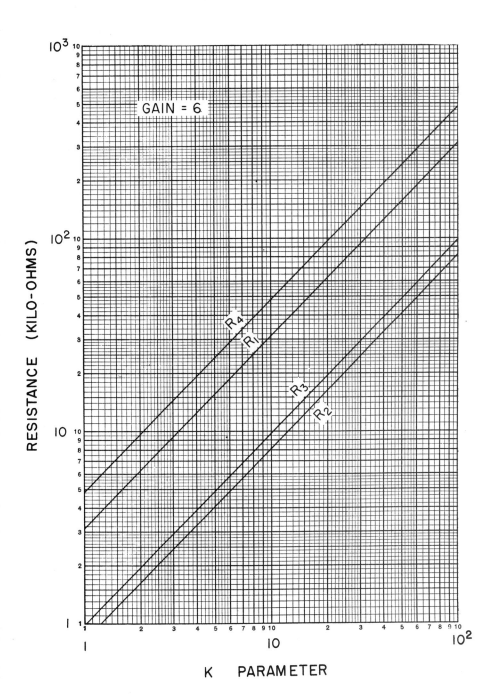

Fig. 3.10  Second-order high-pass Butterworth filter.

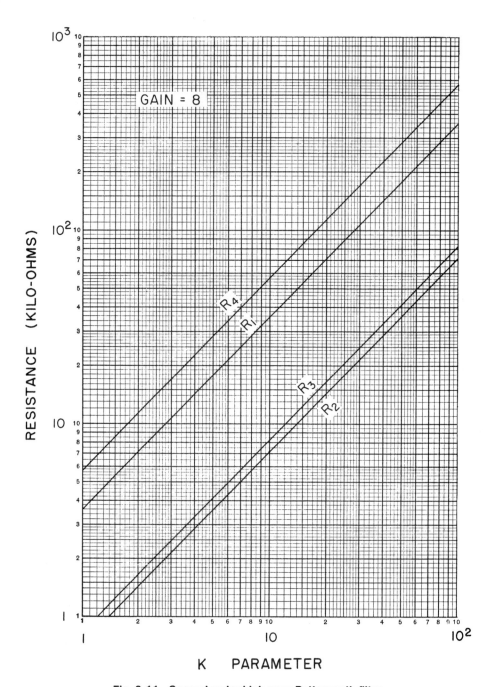

RESISTANCE (KILO-OHMS)

GAIN = 8

$R_4$

$R_1$

$R_3$

$R_2$

K    PARAMETER

Fig. 3.11  Second-order high-pass Butterworth filter.

92

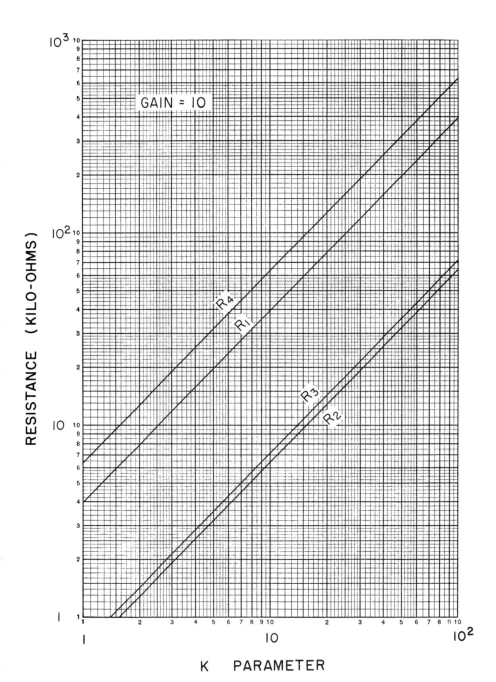

Fig. 3.12 Second-order high-pass Butterworth filter.

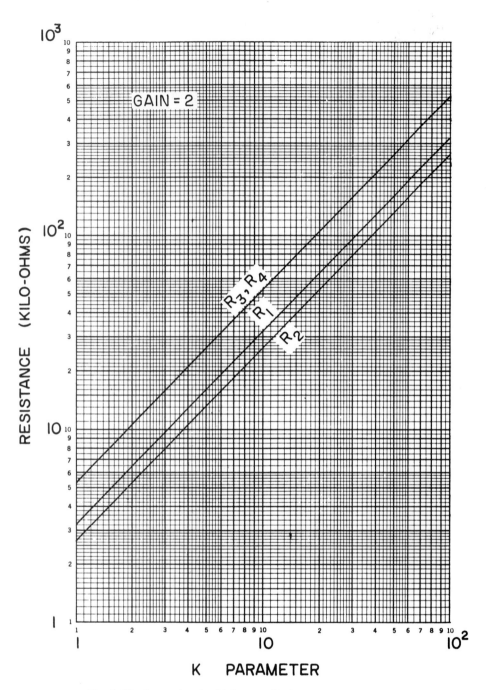

RESISTANCE (KILO-OHMS)

GAIN = 2

$R_3, R_4$

$R_1$

$R_2$

K  PARAMETER

Fig. 3.13  Second-order high-pass Chebyshev filter ($\frac{1}{10}$ dB).

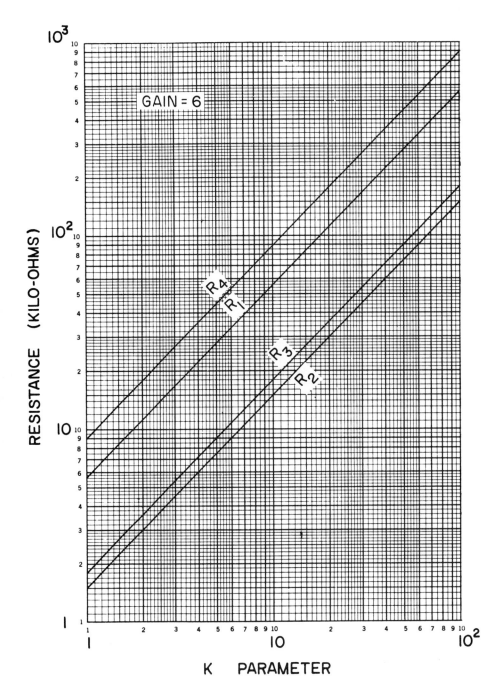

Fig. 3.14  Second-order high-pass Chebyshev filter ( ¹⁄₁₀ dB).

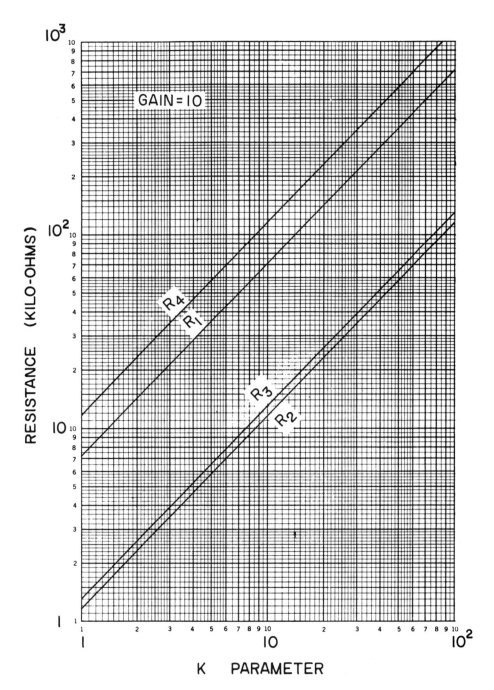

Fig. 3.15  Second-order high-pass Chebyshev filter ( 1/10 dB).

96

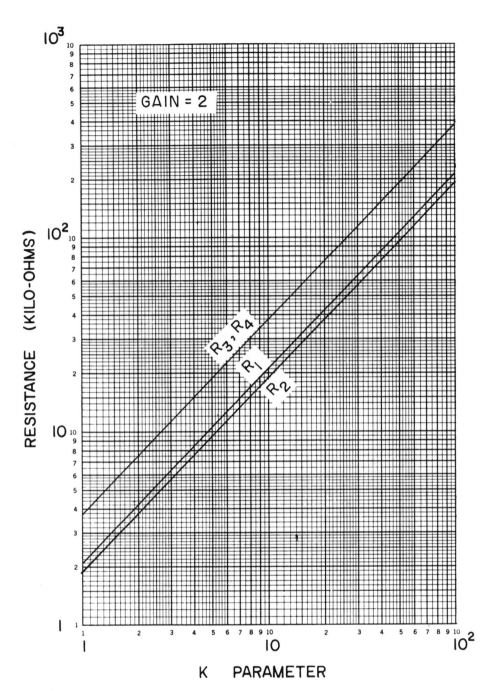

Fig. 3.16 Second-order high-pass Chebyshev filter (½ dB).

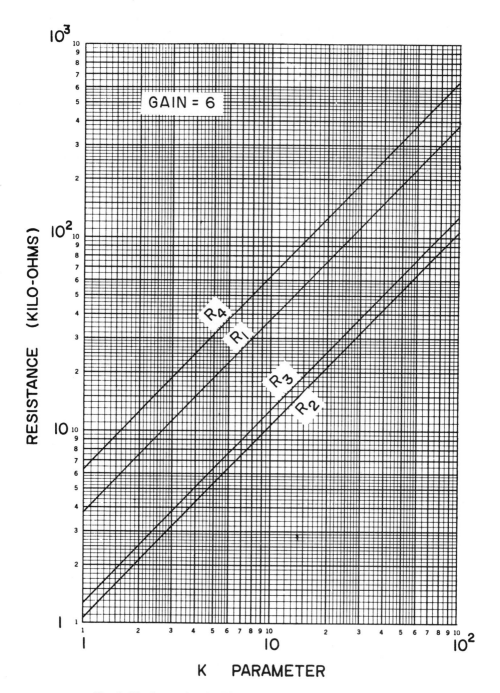

**Fig. 3.17** Second-order high-pass Chebyshev filter ( ½ dB).

98

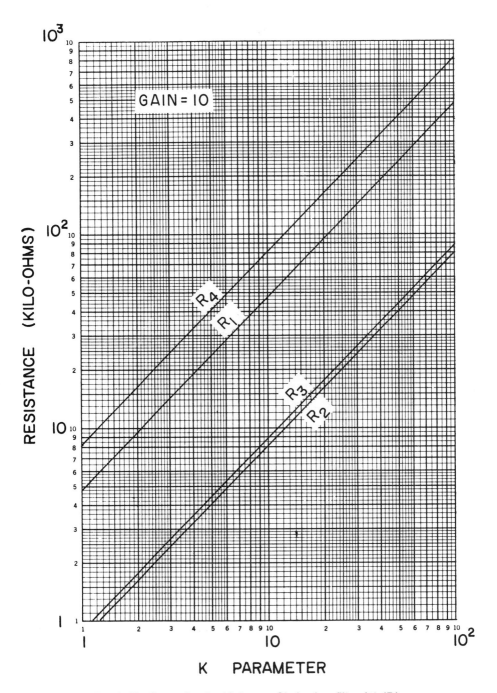

Fig. 3.18   Second-order high-pass Chebyshev filter ( ½ dB ).

99

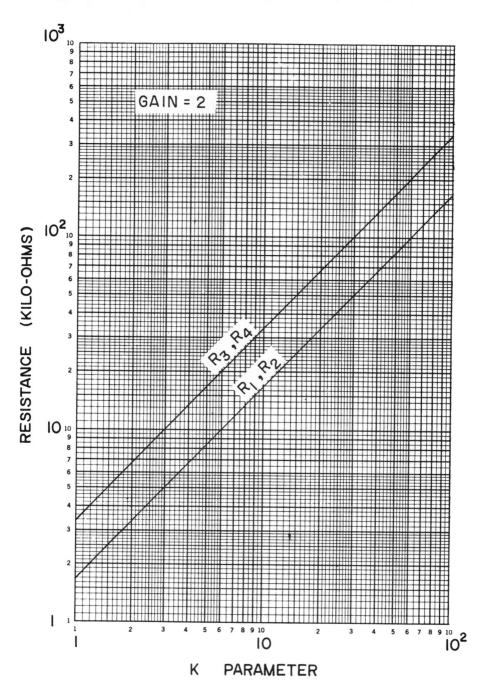

Fig. 3.19  Second-order high-pass Chebyshev filter ( 1 dB ).

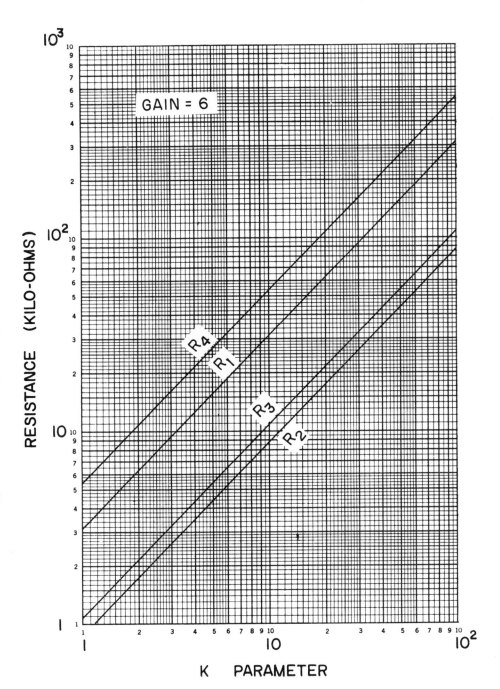

Fig. 3.20   Second-order high-pass Chebyshev filter ( 1 dB).

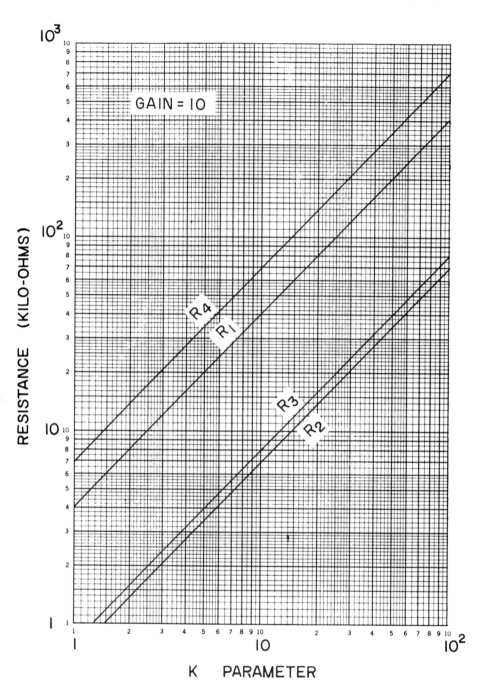

Fig. 3.21 Second-order high-pass Chebyshev filter (1 dB).

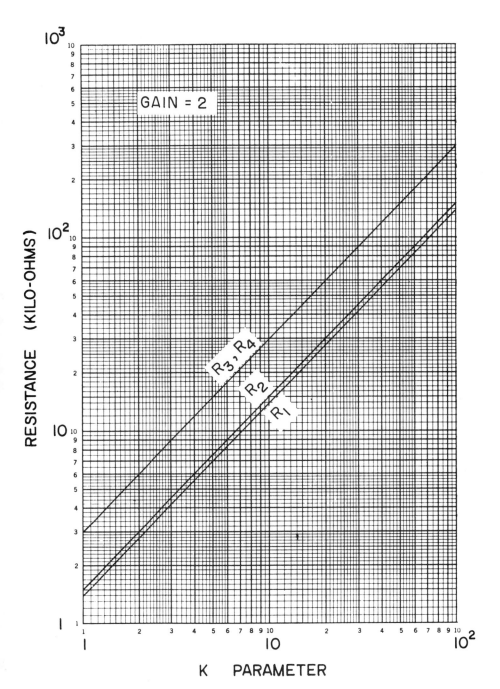

Fig. 3.22 Second-order high-pass Chebyshev filter ( 2 dB ).

103

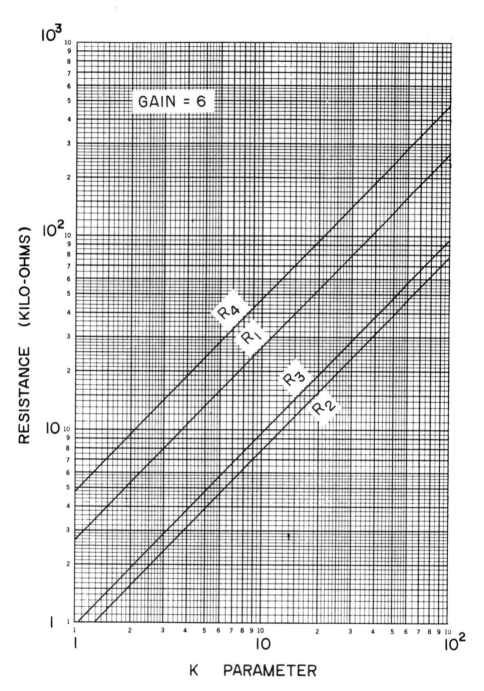

Fig. 3.23 Second-order high-pass Chebyshev filter ( 2 dB).

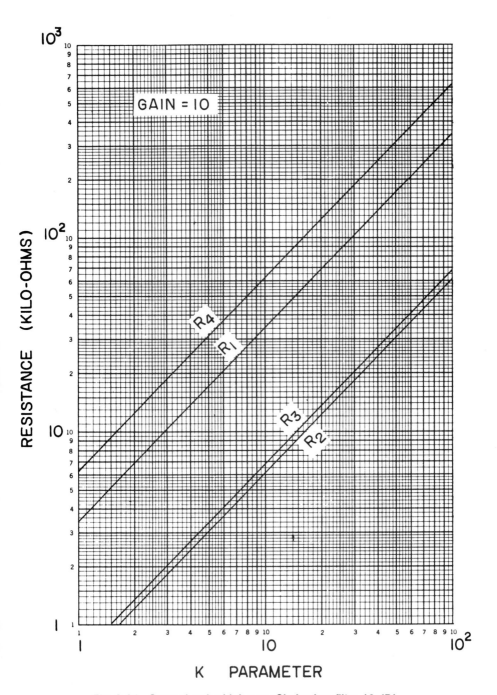

Fig. 3.24  Second-order high-pass Chebyshev filter (2 dB).

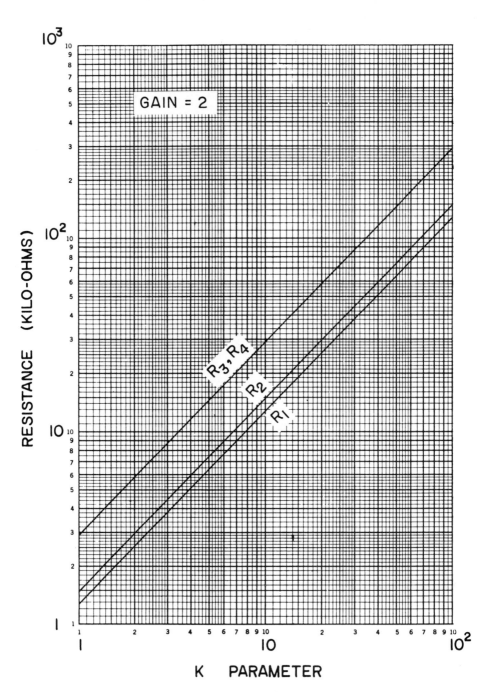

Fig. 3.25   Second-order high-pass Chebyshev filter (3 dB).

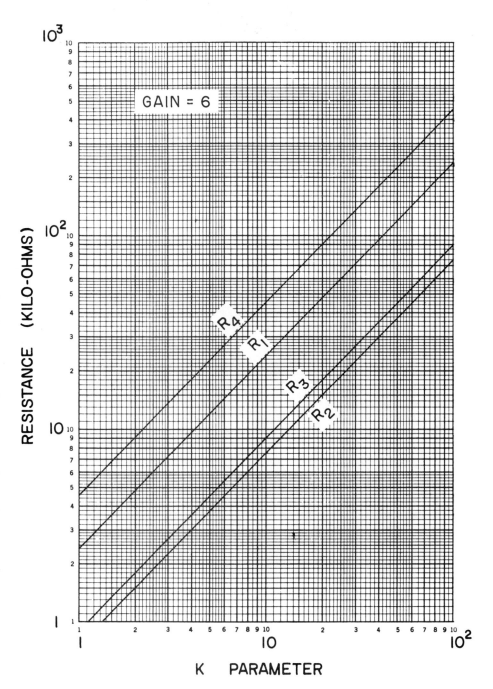

Fig. 3.26 Second-order high-pass Chebyshev filter (3 dB).

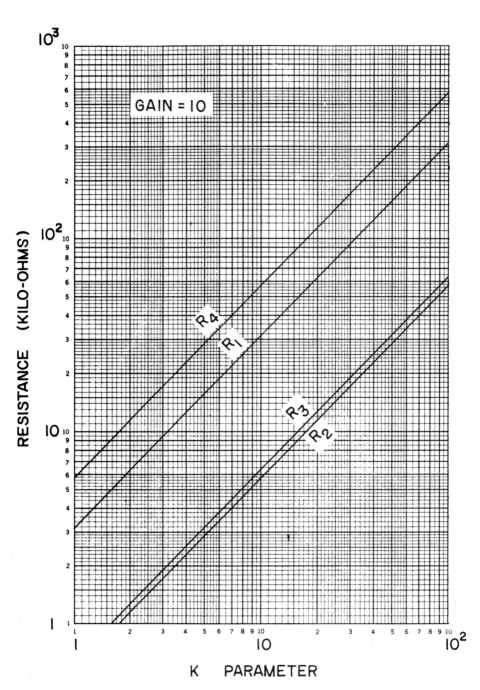

**Fig. 3.27 Second-order high-pass Chebyshev filter ( 3 dB ).**

108

## SUMMARY OF HIGH-PASS FOURTH-ORDER
## FILTER DESIGN PROCEDURE

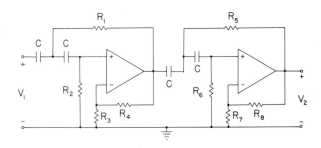

**General circuit.**

### Procedure

Given $f_c$ (Hz), gain, and filter type (Butterworth or Chebyshev), perform the following steps:

1. Select a value of capacitance $C$, determining a $K$ parameter from Fig. 3.6$a$, $b$, or $c$, as described in the second-order high-pass case.
2. Using this value of $K$, find the resistance values of the circuit from the appropriate one of Figs. 3.28 through 3.30 in the Butterworth case and Figs. 3.31 through 3.45 in the Chebyshev case, depending on the gain and, in the Chebyshev case, the dB ripple desired.
3. Select standard resistance values which are as close as possible to those indicated on the graph and construct the circuit.

### Comments and Suggestions

The suggestions given in the second-order low-pass case apply with two exceptions. The dc return to ground is already satisfied by the resistor $R_2$, and the open-loop gain of the op-amp should be at least 50 times the square root of the filter gain.

The remarks in the second-order case for $R_3$ and $R_4$ apply also to $R_7$ and $R_8$.

A specific example is given in Sec. 3.3.

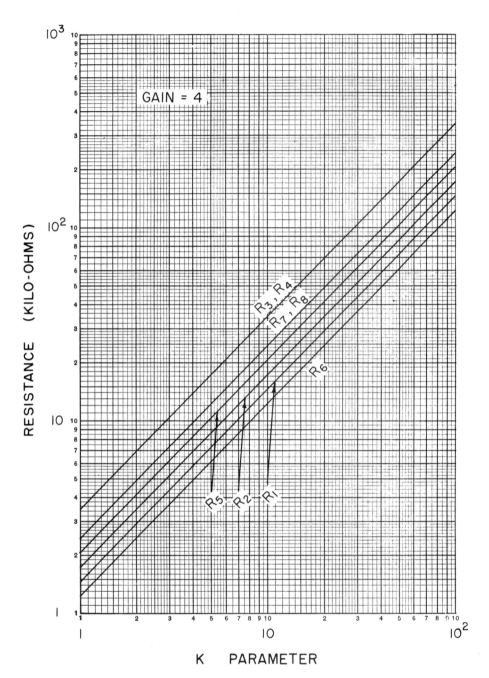

Fig. 3.28 Fourth-order high-pass Butterworth filter.

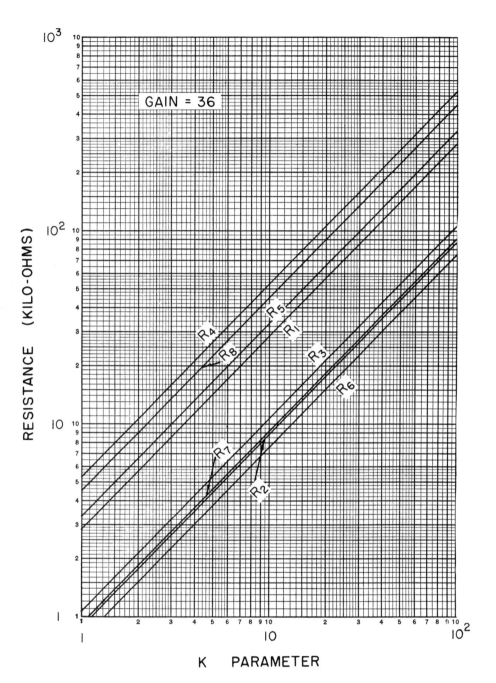

Fig. 3.29 Fourth-order high-pass Butterworth filter.

111

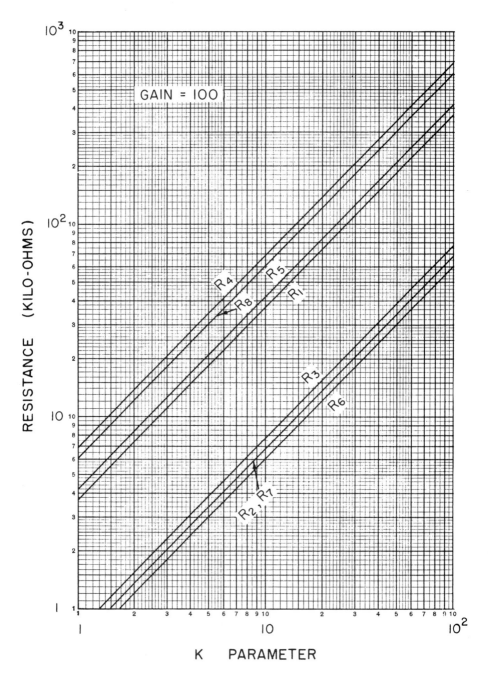

Fig. 3.30  Fourth-order high-pass Butterworth filter.

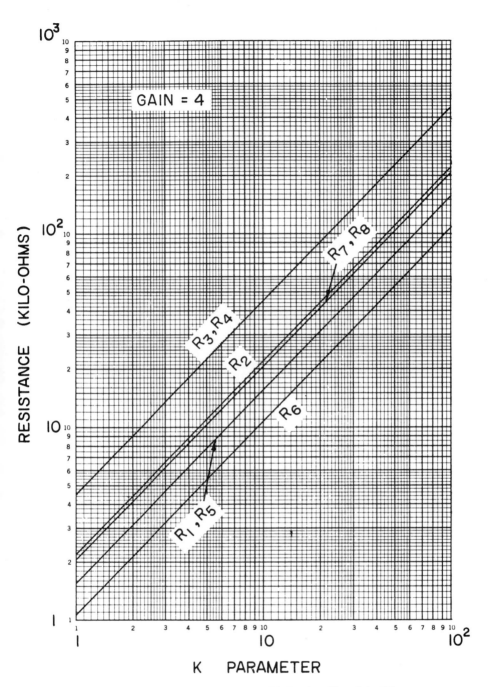

Fig. 3.31　Fourth-order high-pass Chebyshev filter ($\frac{1}{10}$ dB).

113

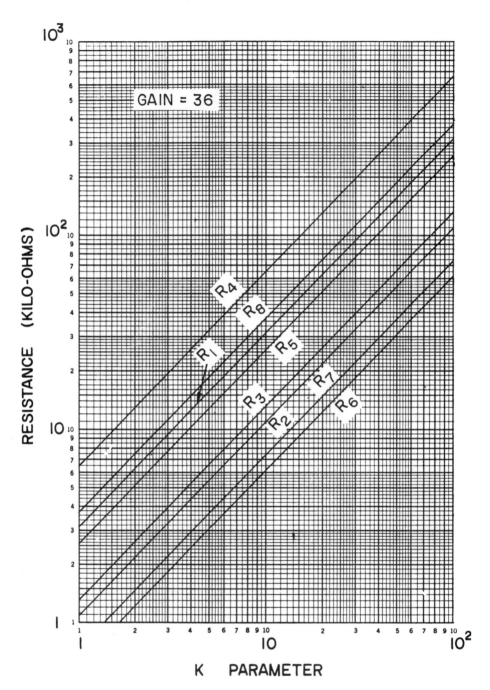

GAIN = 36

Fig. 3.32 Fourth-order high-pass Chebyshev filter ($\frac{1}{10}$ dB).

114

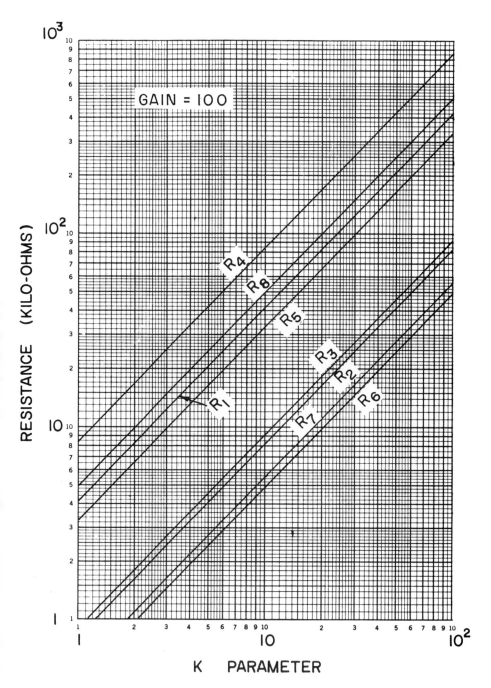

**Fig. 3.33  Fourth-order high-pass Chebyshev filter ( ⅟₁₀ dB ).**

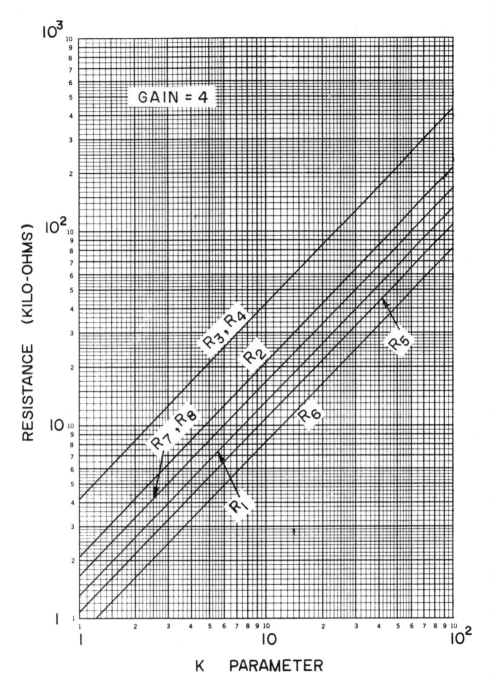

**Fig. 3.34 Fourth-order high-pass Chebyshev filter ( ½ dB ).**

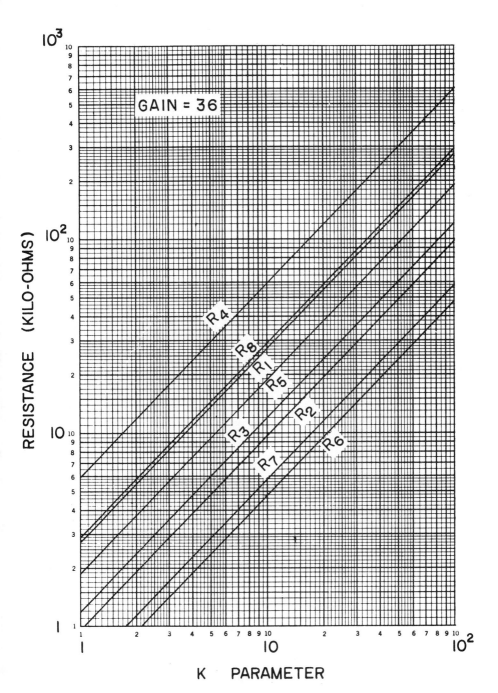

**Fig. 3.35 Fourth-order high-pass Chebyshev filter ( ½ dB).**

117

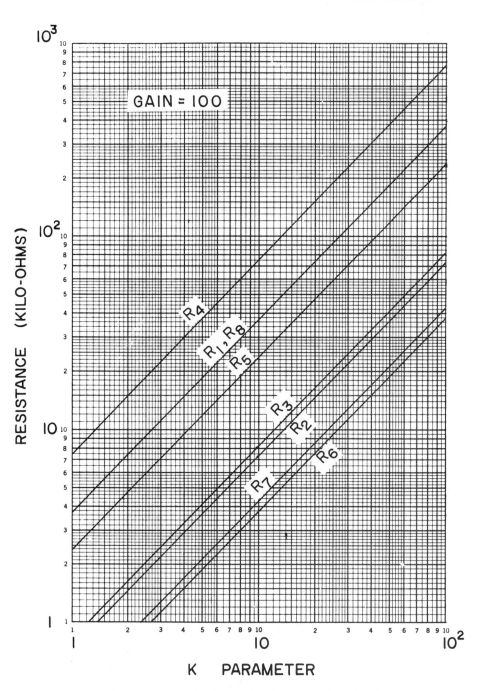

**Fig. 3.36  Fourth-order high-pass Chebyshev filter ( ½ dB).**

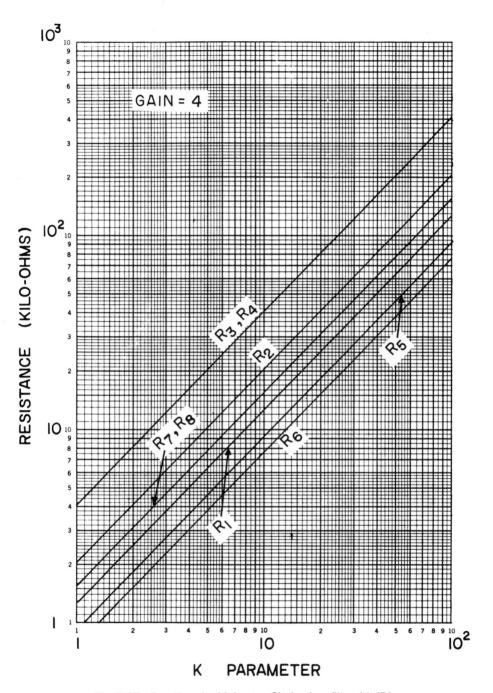

**Fig. 3.37 Fourth-order high-pass Chebyshev filter ( 1 dB).**

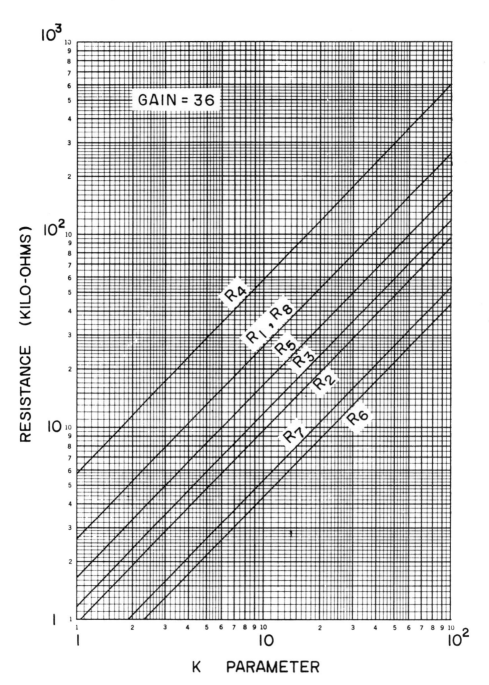

Fig. 3.38  Fourth-order high-pass Chebyshev filter (1 dB).

120

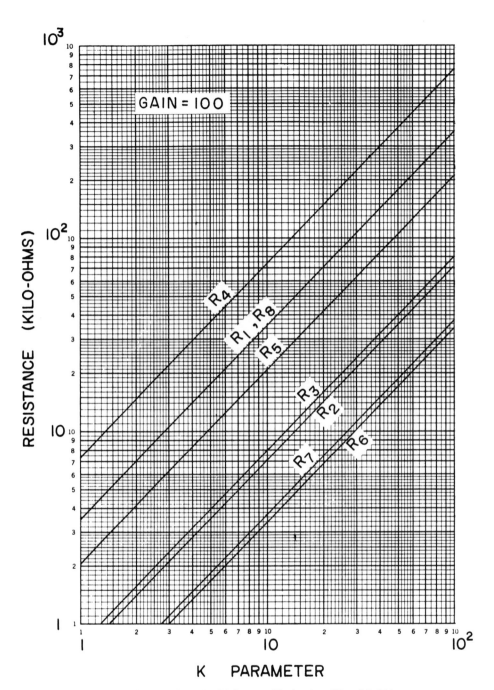

GAIN = 100

RESISTANCE (KILO-OHMS)

K PARAMETER

**Fig. 3.39 Fourth-order high-pass Chebyshev filter ( 1 dB).**

121

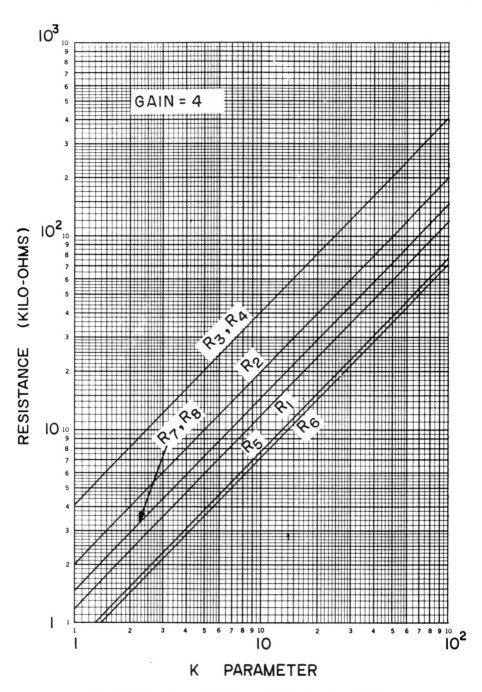

Fig. 3.40 Fourth-order high-pass Chebyshev filter ( 2 dB ).

122

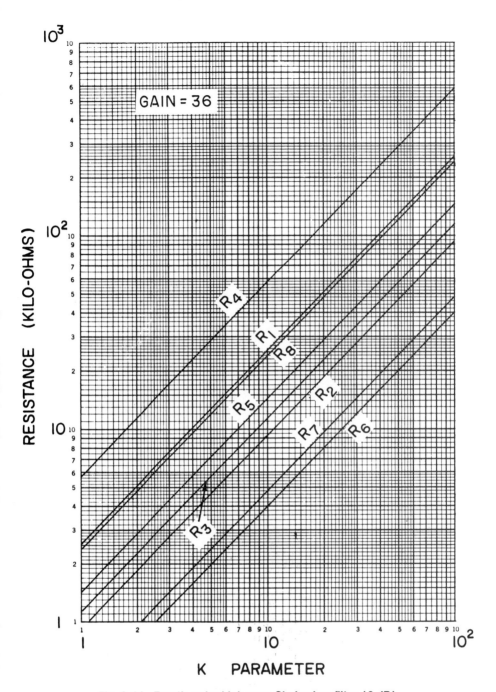

Fig. 3.41  Fourth-order high-pass Chebyshev filter ( 2 dB).

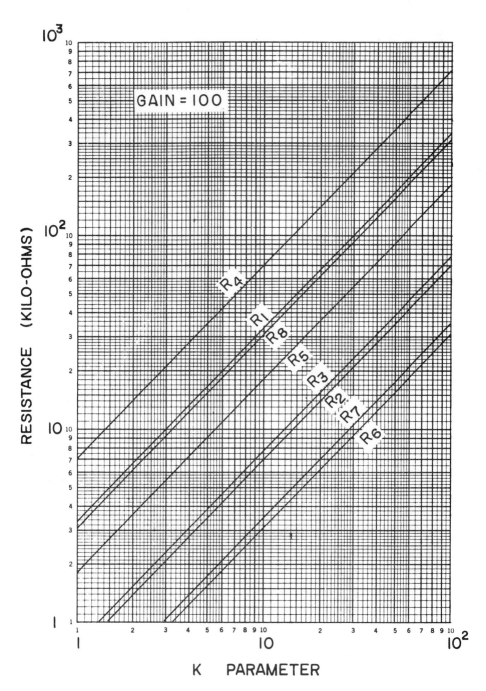

RESISTANCE (KILO-OHMS)

GAIN = 100

$R_4$
$R_1$
$R_8$
$R_5$
$R_3$
$R_2$
$R_7$
$R_6$

K   PARAMETER

Fig. 3.42   Fourth-order high-pass Chebyshev filter (2 dB).

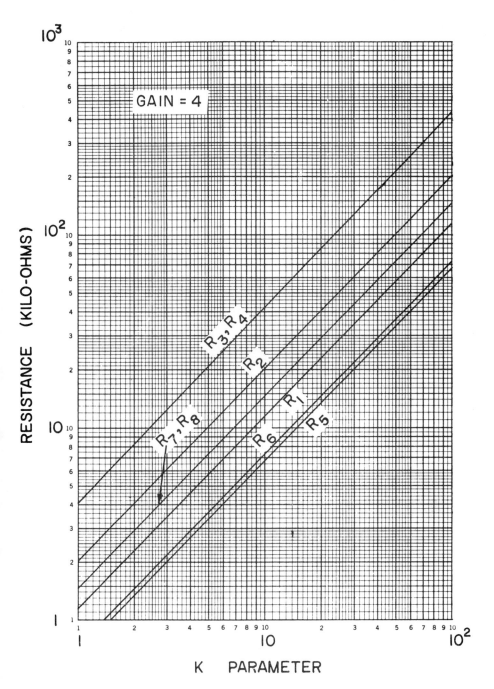

Fig. 3.43  Fourth-order high-pass Chebyshev filter ( 3 dB ).

125

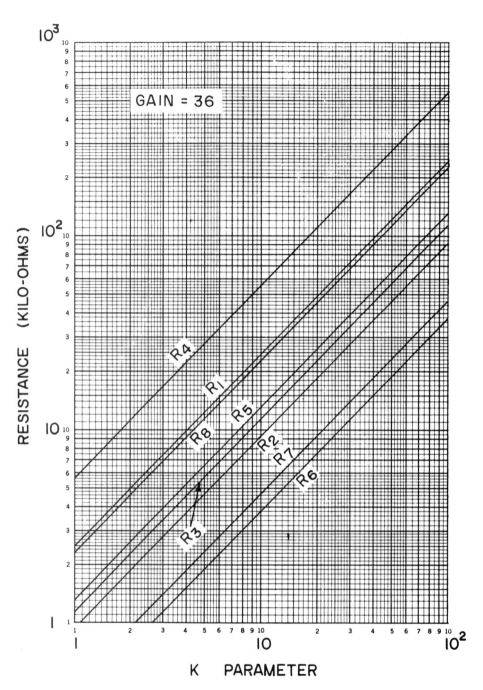

Fig. 3.44 Fourth-order high-pass Chebyshev filter (3 dB).

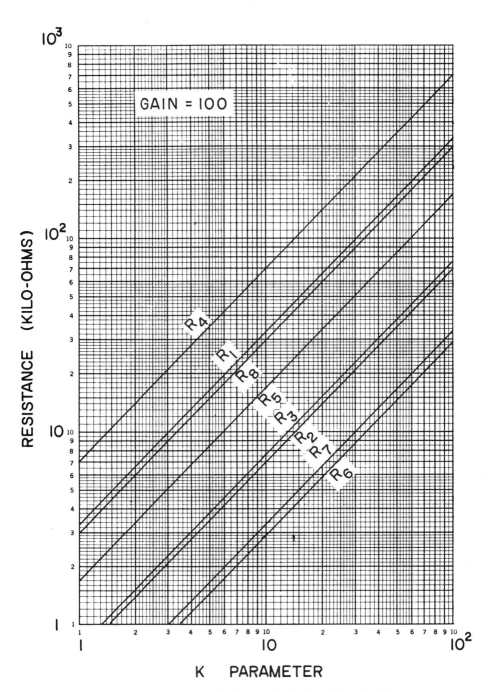

Fig. 3.45 Fourth-order high-pass Chebyshev filter (3 dB).

# 4 BAND-PASS FILTERS

## 4.1 General Circuit and Equations: Second-Order Case

A band-pass filter passes a band of frequencies of bandwidth $B$ centered approximately about a center frequency $\omega_0$, and attenuates all other frequencies. Both $B$ and $\omega_0$ may be measured in radians per second, or $B$ may be given in Hz with a center frequency $f_0 = \omega_0/2\pi$ Hz. An ideal band-pass response is that represented by the broken line, and an approximation to the ideal is that represented by the solid line in Fig. 4.1. A second-order approximation to the ideal is achieved, for appropriate values of $B$ (rad/sec) and $\omega_0^2$, by the transfer function [21]

$$H(s) = \frac{V_2(s)}{V_1(s)} = \frac{Ks}{s^2 + Bs + \omega_0^2} \qquad (4.1)$$

Another quantity of interest in a band-pass filter is the quality factor $Q$, defined by $Q = \omega_0/B$, or if $B$ is in Hz, $Q = f_0/B$. Hence a high $Q$ indicates a highly selective filter since the band of frequencies which pass is narrow

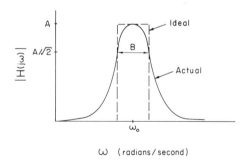

Fig. 4.1   A band-pass amplitude response.

compared to the center frequency. The gain of the filter is defined as the amplitude of $H(s)$ at the center frequency, and from Eq. (4.1), is seen to be gain $= K/B$.

### 4.2 A Second-Order VCVS Band-Pass Filter

A circuit which realizes Eq. (4.1), and one of which we shall use to obtain second-order band-pass filters, is the VCVS circuit of Fig. 4.2, from Kerwin and Huelsman [25]. Analysis of the circuit shows that Eq. (4.1) is realized if

$$K = \frac{\mu}{R_1 C}$$

$$B = \frac{4 - \mu}{R_1 C} \tag{4.2}$$

$$\omega_0^2 = \frac{2}{R_1^2 C^2}$$

where

$$\mu = 1 + \frac{R_3}{R_2} \tag{4.3}$$

The circuit of Fig. 4.2 performs best for low values of $Q$, as may be seen from Eqs. (4.2). Since $4 - \mu = R_1 C \omega_0 / Q$, we note that for high values of $Q$, $\mu$ is very near 4 and hence from Eqs. (4.2), is quite sensitive to changes in $R_2$ and $R_3$. For this reason, the curves we give for practical usage of Fig. 4.2 are limited to $Q \leq 4$. However, as described in the summary at the end of this chapter, higher $Q$s are possible using a potentiometer. A nice feature of Fig. 4.2 is that the center frequency $f_0$ and the chosen value of $C$ determine $R_1$. Thus we may obtain a variety of bandwidths for a fixed center frequency by changing only $R_2$ and $R_3$.

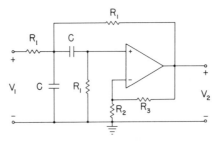

**Fig. 4.2   A second-order VCVS band-pass filter.**

A practical design of the form of Fig. 4.2 may be obtained for a given $f_0$ and $B$ (in Hz), or equivalently a given $f_0$ and $Q$, by using the procedure described in the summary at the end of the chapter.

For example, suppose $f_0 = 20,000$ Hz and $Q = 2$ ($B = f_0/Q = 10,000$ Hz). From Fig. 4.12c, for $C = 0.001$ $\mu$F, we have $K$ parameter $= 5$, and from Fig. 4.13 we have $R_1 = 11.2$ k$\Omega$, $R_2 = 16.2$ k$\Omega$, $R_3 = 32$ k$\Omega$, and gain $= 4.7$. Using the standard resistances 11, 15, and 33 k$\Omega$, and a $\mu$A748 op-amp with a compensation of 10 pF, we have a filter whose amplitude response is shown in Fig. 4.3. The scale shown is 5000 Hz/division. The actual results were $f_0 = 19,400$ Hz, $Q = 2$, and gain $= 4.35$.

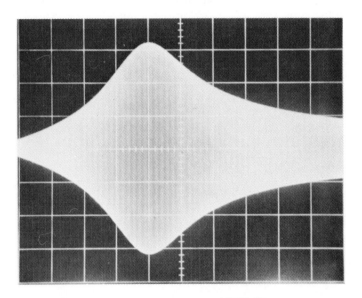

**Fig. 4.3  A second-order band-pass VCVS filter response.**

If we desire a different gain from the one we get from Eqs. (4.2), we may use a VCVS preamplifier, as discussed in Sec. 1.4, and shown in Fig. 1.2. For instance, in the example just considered, suppose we want a gain of 10. We have a gain of 4.7, so that the preamplifier must supply a gain of $10/4.7 = 2.13$. By Eq. (1.2) we must have

$$\frac{R_a}{R_b} = 2.13 - 1 = 1.13$$

Therefore if we choose $R_b = 1.6$ k$\Omega$, then $R_a = 1.13 \times 1.6 = 1.81$ k$\Omega$, so that we may use the standard values of 1.6 and 1.8 k$\Omega$.

### 4.3   A Second-Order Multiple-Feedback Band-Pass Filter

Another circuit which realizes the second-order band-pass filter is the multiple-feedback network [26] shown in Fig. 4.4. Analysis shows that it realizes Eq. (4.1) for the values

$$B = \frac{2}{R_3 C}$$

$$\omega_0^2 = \frac{1}{R_3 C^2}\left(\frac{1}{R_1} + \frac{1}{R_2}\right)$$

(4.4)

The constant $K$ in Eq. (4.1) is given by $-1/R_1 C$, and hence the circuit yields an inverting gain (negative) with magnitude $R_3/2R_1$. This may be converted to a noninverting gain, if one wishes, by cascading Fig. 4.4 with an inverting amplifier.

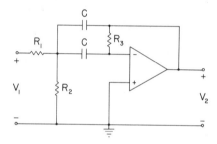

**Fig. 4.4   A multiple-feedback second-order band-pass filter.**

For high $Q$, the network of Fig. 4.4 has a wide spread of element values and large $Q$ sensitivities. For this reason it should probably be restricted to values of $Q \le 10$. The network has the nice feature that one may specify $f_0$, $Q$, and the gain. Practical values of the capacitances and resistances may be obtained using the procedure described in the summary at the end of the chapter.

As an example, let $f_0 = 1000$ Hz, $Q = 10$, and gain $= 10$. From Fig. 4.12b, we have a $K$ parameter of 10, and from Fig. 4.32, we have $R_1 = 15.9$ kΩ, $R_2 = 840$ Ω, $R_3 = 318$ kΩ, for a $C$ of 0.01 μF. Using an SU536 op-amp and resistances of 16 kΩ, 820 Ω, and 330 kΩ respectively, we obtain the response shown in Fig. 4.5, having $f_0 = 1024$ Hz, $Q = 9.3$ ($B = 110$ Hz), and a gain of 8.8. The response is shown with a scale of 250 Hz/division.

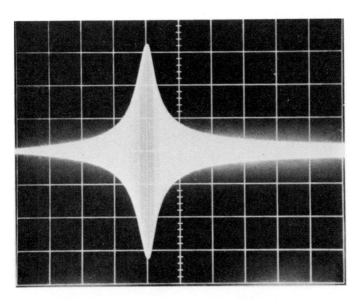

**Fig. 4.5    A multiple-feedback band-pass filter response.**

## 4.4    A Second-Order Positive-Feedback Band-Pass Filter

The second-order band-pass filters of the previous two sections are limited, for best results, to $Q$s of 10 or so. A circuit using two op-amps, which may be used to obtain values of $Q$ up to 50, is the positive feedback circuit [2] of Fig. 4.6. (Positive feedback means that the signal fed back, in this case through $R_3$, is a noninverted signal.)

Analysis of Fig. 4.6 shows that Eq. (4.1) is satisfied for the values

$$K = \frac{R_4}{R_1^2 C}$$

$$B = \frac{1}{R_1 C} \left( 2 - \frac{R_4}{R_3} \right)$$

$$\omega_0^2 = \frac{1}{R_1 C^2} \left( \frac{1}{R_1} + \frac{1}{R_2} + \frac{1}{R_3} \right)$$

(4.5)

A practical design of Fig. 4.6 for a given $f_0$, $Q$, and $C$ may be obtained as described in the summary.

As an example, let $f_0 = 2000$ Hz, gain $= 4$, $Q = 40$, and select $C = 0.01 \ \mu\text{F}$. From Fig. 4.12$b$, we have a $K$ parameter of 5, and from Fig. 4.36,

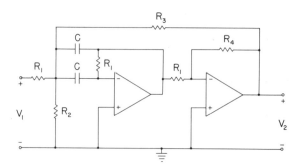

**Fig. 4.6    A second-order band-pass positive-feedback filter.**

we have $R_1 = 50.5$ kΩ, $R_2 = 1.38$ kΩ, $R_3 = 17.3$ kΩ, and $R_4 = 31.5$ kΩ. Using the values 51, 1.3, 18, and 33 kΩ respectively and a CA3056A op-amp, we obtain the response shown in Fig. 4.7, with $f_0 = 2037$ Hz, $Q = 43.3$, and a gain of 4.4. The picture was taken with 500 Hz/division as the scale.

The quality factor $Q$ (and hence the bandwidth) can be varied to some degree, without appreciably changing $f_0$, by varying $R_3$ or $R_4$. For example, by varying $R_4$ with a potentiometer to a value of 36.6 kΩ, we obtained a $Q$

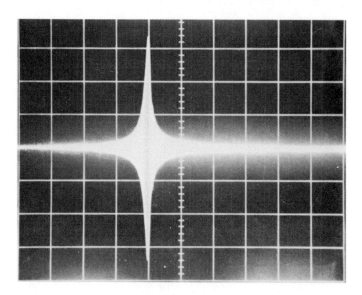

**Fig. 4.7    A positive-feedback band-pass filter response.**

of 102, a gain of 12.5, and changed $f_0$ only slightly to 2035 Hz. To vary $f_0$ back to 2000 Hz, one could increase $C$ slightly by paralleling with it a small capacitance.

## 4.5  A Biquad Band-Pass Filter

A circuit using three op-amps which can be used to obtain values of $Q$ up to 100 is the *biquad* band-pass filter [18] of Fig. 4.8. Analysis shows that it realizes Eq. (4.1) for the values

$$K = \frac{1}{R_1 C}$$

$$B = \frac{1}{R_2 C}$$

$$\omega_0^2 = \frac{1}{R_3 R_4 C^2}$$

The gain $K/B$ is therefore $R_2/R_1$, and if an inverting (negative) gain is desired, the output may be taken at node $a$.

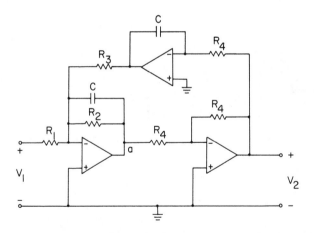

**Fig. 4.8  A second-order biquad band-pass filter.**

The biquad circuit has more elements than the other band-pass circuits, but it can realize a much higher $Q$ and is particularly easy to adjust. The gain is adjusted by varying $R_1$, $Q$ is adjusted by varying $R_2$, and $f_0$ is adjusted by changing $R_3$. The design graph for the biquad filter is Fig. 4.41.

To illustrate the use of the biquad circuit, let us solve the example considered previously in Sec. 4.4. The center frequency was $f_0 = 2000$ Hz, the gain was 4, and $Q$ was 40. For $C = 0.01$ $\mu$F, the $K$ parameter was 5. From Fig. 4.41 with $K = 5$, we have $R_3 = R_4 = 8$ k$\Omega$, $R_2 = R_3 Q = 8 \times 40 = 320$ k$\Omega$, and $R_1 = R_2/\text{GAIN} = 320/4 = 80$ k$\Omega$.

### 4.6 Higher Order Band-Pass Filters

We may obtain a fourth-order band-pass filter transfer function by multiplying two functions of the type given in Eq. (4.1), resulting in a constant times $s^2$ divided by a fourth-degree polynomial. Such a function may be interpreted as the product of a low-pass second-order transfer function and a high-pass second-order transfer function. Thus a quick way to get a variety of fourth-order band-pass filters is to cascade a low-pass second-order filter with cutoff point $f_{c2}$ with a high-pass second-order filter with cutoff point $f_{c1} < f_{c2}$, making use of the results obtained in Chapters 2 and 3. The result is a filter with a center frequency approximately $f_0 = \sqrt{f_{c1}f_{c2}}$ (it is exactly this when both filters are Butterworth filters) and $B$ approximately equal to $f_{c2} - f_{c1}$. The approximations improve as the difference $f_{c2} - f_{c1}$ increases, and hence as $Q$ decreases. Best results are obtained if $f_{c1}$ and $f_{c2}$ are at least one octave apart [27], and hence for $Q$ not exceeding approxi-

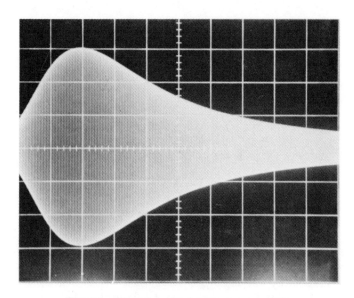

Fig. 4.9 A fourth-order band-pass response.

mately $\sqrt{2} = 1.414$. (An octave is the interval between two frequencies, one twice the other.) Sharper cutoff features may be obtained by cascading fourth-order low- and high-pass filters, of course, but this requires four op-amps.

In the case where both the high- and low-pass filters are Butterworth filters, if $f_{c1} = f_{c2}$, then the common value is also $f_0$ and $Q$ is 1.1. If $f_{c2} = 2f_{c1}$ (one octave apart), then $f_0 = \sqrt{2}f_{c1} = 1.414f_{c1}$ and $Q = 0.912$.

As an example, let us cascade a second-order Butterworth low-pass filter with $f_c = 4000$ Hz and a gain of 4 with a high-pass Butterworth filter of second order with $f_c = 2000$ Hz and a gain of 2. This should result in $f_0$ approximately $\sqrt{2000 \times 4000} = 2828$ Hz and a gain of about $4 \times 2 = 8$. The low-pass filter has element values $R_1 = 1.8$ kΩ, $R_2 = 9.1$ kΩ, $R_3 = 15$ kΩ, $R_4 = 43$ kΩ, and $C = C_1 = 0.01$ µF (refer to Fig. 2.3). The high-pass section has values $R_1 = 9.1$ kΩ, $R_2 = 6.8$ kΩ, $R_3 = R_4 = 15$ kΩ, and $C = 0.01$ µF (refer to Fig. 3.2). Using a 747 op-amp, the results are $f_0 = 2820$ Hz, gain $= 6.9$, and $Q = 0.95$. The response is shown in Fig. 4.9, where the scale is 1000 Hz/division, starting at 1000 Hz.

A much sharper band-pass filter may be obtained by cascading two or more identical band-pass second-order filters. If $Q_1$ is the quality factor of a single stage and there are $n$ stages, then the $Q$ of the filter is $Q_1/\sqrt{\sqrt[n]{2} - 1}$. These values and the corresponding bandwidths are shown, for $n = 1, 2, \ldots, 5$, in the table of Fig. 4.10, where $B_1$ is the bandwidth of a filter with a single stage.

| $n$ | Bandwidth | $Q$ |
|---|---|---|
| 1 | $B_1$ | $Q_1$ |
| 2 | $0.644 B_1$ | $1.55 Q_1$ |
| 3 | $0.510 B_1$ | $1.96 Q_1$ |
| 4 | $0.435 B_1$ | $2.30 Q_1$ |
| 5 | $0.386 B_1$ | $2.60 Q_1$ |

**Fig. 4.10   A table of $Q$ values for cascaded band-pass filters.**

As an example, the band-pass filter of Fig. 4.5 was designed for $f_0 = 1000$ Hz, $Q = 10$, and a gain of 10. The actual results were $f_0 = 1024$ Hz, $Q = 9.3$, and a gain of 8.8. Cascading this filter with another one identical to it results in the picture shown in Fig. 4.11, where $f_0 = 1028$ Hz, $Q = 14.3$, and the gain is 88. The picture has a scale of 250 Hz/division.

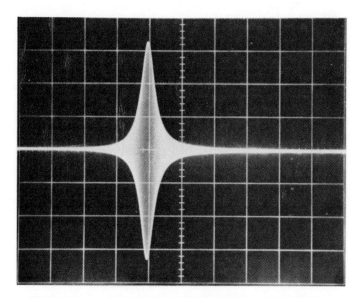

**Fig. 4.11   The response of a filter with two identical stages.**

Summaries of the techniques for obtaining the various practical band-pass filters are given, together with the graphs, following this section.

## SUMMARY OF SECOND-ORDER VCVS BAND-PASS FILTER DESIGN PROCEDURE ( $Q \leq 4$ )

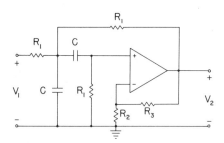

**General circuit.**

## Procedure

Given $f_0$ (Hz), $Q$ (or bandwidth $BW$ in Hz), perform the following steps:

1. Select a value of capacitance $C$, determining a $K$ parameter from Fig. $4.12a$ if $f_0$ is between 1 and $10^2 = 100$, from Fig. $4.12b$ if $f_0$ is between 100 and $10^4 = 10,000$, and from Fig. $4.12c$ if $f_0$ is between 10,000 and $10^6 = 1,000,000$ Hz.
2. Using this value of $K$, find the resistances from the appropriate one of Figs. 4.13 through 4.15, depending on $Q$ (or $BW$).
3. Select standard resistances which are as close as possible to those indicated on the graph and construct the circuit.

## Comments and Suggestions

The remarks given for the second-order VCVS low-pass filter are applicable with the following exceptions:

(1) The statement concerning $R_3$ and $R_4$ applies to $R_2$ and $R_3$.
(2) The dc return to ground is already satisfied by $R_1$.
(3) Remarks concerning $f_c$ now apply to $f_0$.

The center frequency $f_0$ can be fixed and the bandwidth (or $Q$) changed by varying with a potentiometer the ratio $R_3/R_2$. (See Sec. 4.2.)

A specific example is given in Sec. 4.2.

The values of $Q$, bandwidth, and gain, for $N$ identical cascaded sections, $N = 1, 2, 3, 4$, are shown on Figs. 4.13 through 4.15.

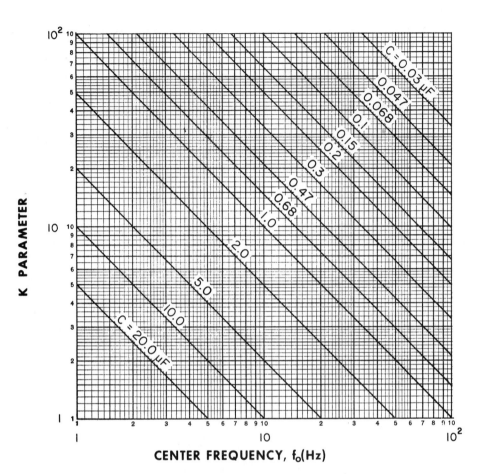

**K PARAMETER**

**CENTER FREQUENCY, f₀(Hz)**

**Fig. 4.12(a)** *K* parameter versus frequency.

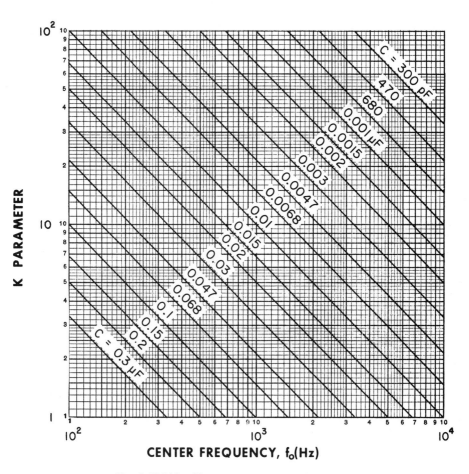

**K PARAMETER**

**CENTER FREQUENCY, f₀(Hz)**

Fig. 4.12(*b*)   *K* parameter versus frequency.

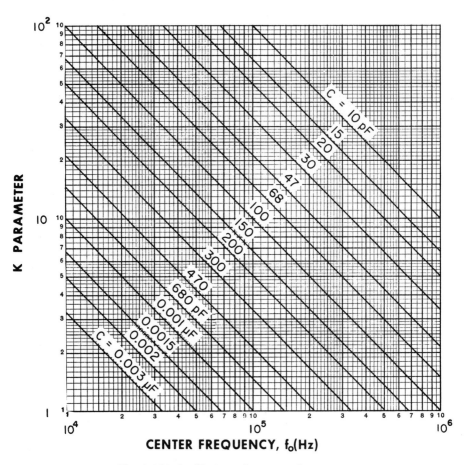

**K PARAMETER**

**CENTER FREQUENCY, f₀(Hz)**

**Fig. 4.12 ( c )** *K* parameter versus frequency.

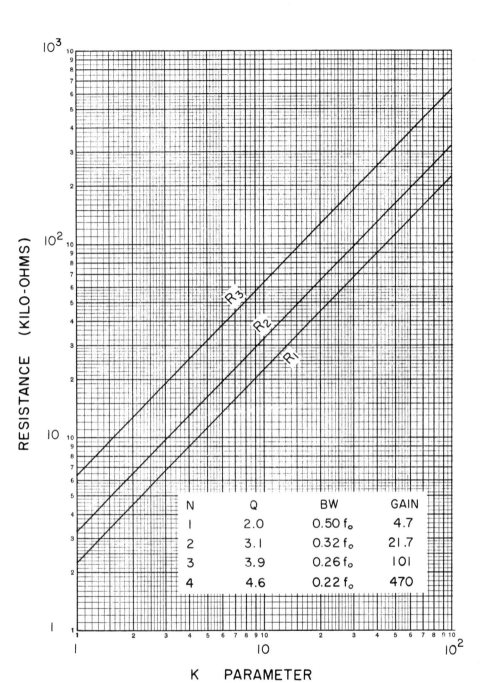

| N | Q | BW | GAIN |
|---|-----|----------|------|
| 1 | 2.0 | 0.50 $f_o$ | 4.7 |
| 2 | 3.1 | 0.32 $f_o$ | 21.7 |
| 3 | 3.9 | 0.26 $f_o$ | 101 |
| 4 | 4.6 | 0.22 $f_o$ | 470 |

**K  PARAMETER**

**Fig. 4.13  VCVS band-pass filter.**

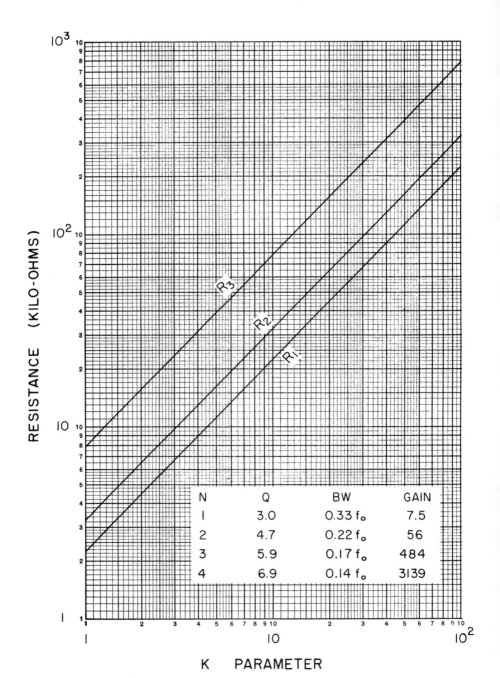

| N | Q | BW | GAIN |
| 1 | 3.0 | 0.33 f$_o$ | 7.5 |
| 2 | 4.7 | 0.22 f$_o$ | 56 |
| 3 | 5.9 | 0.17 f$_o$ | 484 |
| 4 | 6.9 | 0.14 f$_o$ | 3139 |

**K    PARAMETER**

**Fig. 4.14   VCVS band-pass filter.**

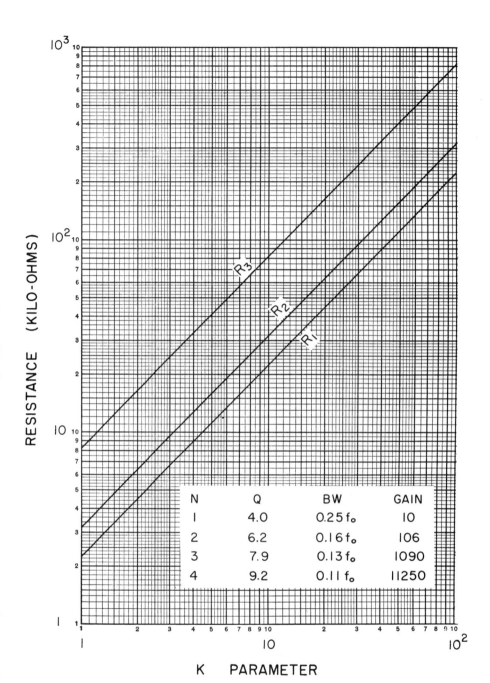

| N | Q | BW | GAIN |
|---|-----|---------|-------|
| 1 | 4.0 | $0.25 f_o$ | 10 |
| 2 | 6.2 | $0.16 f_o$ | 106 |
| 3 | 7.9 | $0.13 f_o$ | 1090 |
| 4 | 9.2 | $0.11 f_o$ | 11250 |

**RESISTANCE (KILO-OHMS)**

**K PARAMETER**

**Fig. 4.15 VCVS band-pass filter.**

## SUMMARY OF SECOND-ORDER MULTIPLE-FEEDBACK BAND-PASS FILTER DESIGN PROCEDURE ( $Q \leq 10$ )

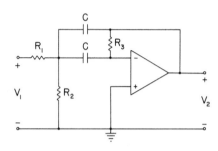

**General circuit.**

### Procedure

Given $f_0$ (Hz), $Q$ (or bandwidth $BW$ in Hz), and gain, perform the following steps:

1. Select a value of capacitance $C$ and determine a $K$ parameter from Fig. 4.12$a$, $b$, or $c$, as described for the second-order VCVS band-pass filter.
2. Using this value of $K$, find the resistances from the appropriate one of Figs. 4.16 to 4.32, depending on $Q$ (or $BW$) and gain.
3. Select standard resistances which are as close as possible to those indicated on the graph and construct the circuit.

### Comments and Suggestions

The remarks given for the second-order low-pass filter are applicable with the following exceptions:

(1) The statement concerning the ratio $R_4/R_3$ is not applicable.
(2) The dc return to ground is already satisfied by $R_3$.
(3) Remarks concerning $f_c$ now apply to $f_0$.

For tuning purposes to a limited degree, varying $R_3$ with a potentiometer affects $Q$ and hence $BW$, whereas varying $C$ changes $f_0$. For minimum dc offset, a resistance equal to $R_3$ can be placed in the noninverting input to ground.

A specific example is given in Sec. 4.3.

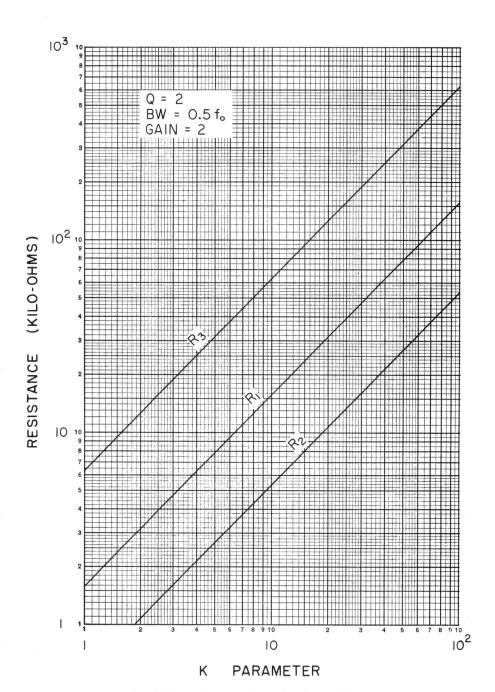

RESISTANCE (KILO-OHMS)

Q = 2
BW = 0.5 f$_o$
GAIN = 2

R$_3$

R$_1$

R$_2$

K    PARAMETER

**Fig. 4.16  Multiple-feedback band-pass filter.**

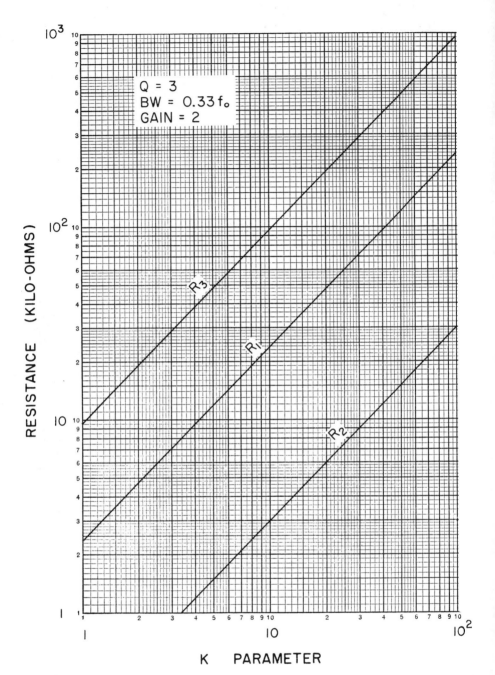

Q = 3
BW = 0.33 f₀
GAIN = 2

**Fig. 4.17 Multiple-feedback band-pass filter.**

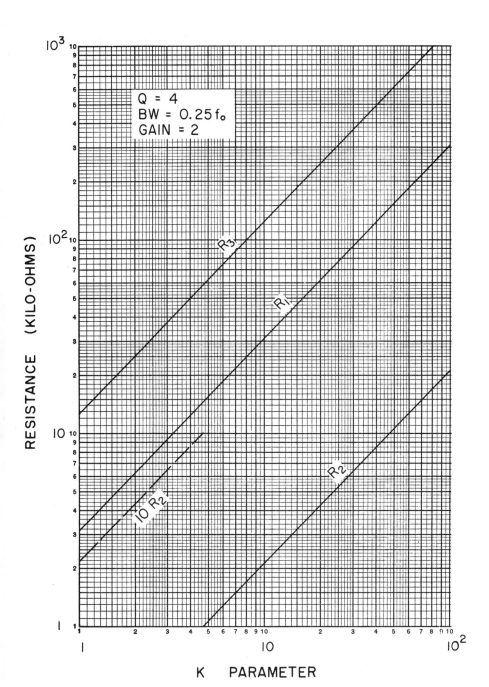

Fig. 4.18 Multiple-feedback band-pass filter.

149

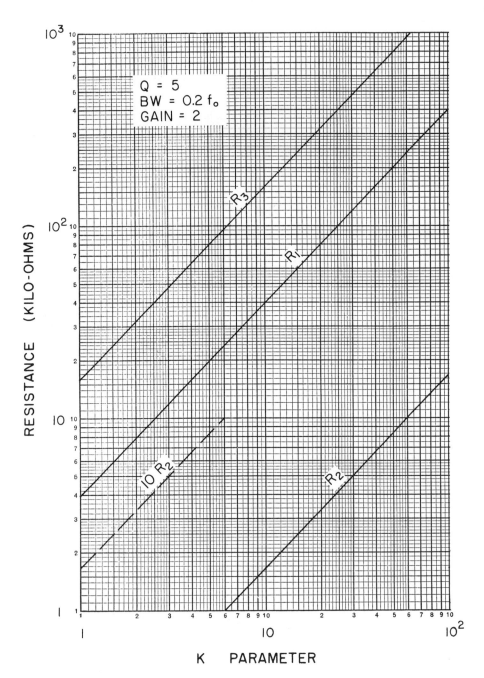

Fig. 4.19  Multiple-feedback band-pass filter.

Q = 5
BW = 0.2 f₀
GAIN = 2

RESISTANCE (KILO-OHMS)

K   PARAMETER

150

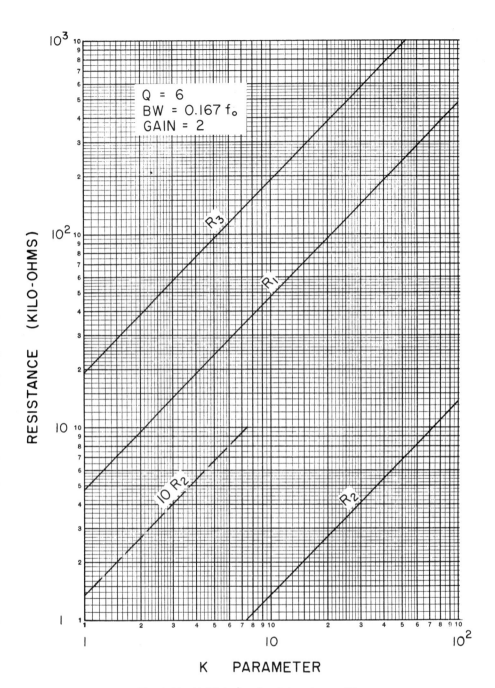

Fig. 4.20   Multiple-feedback band-pass filter.

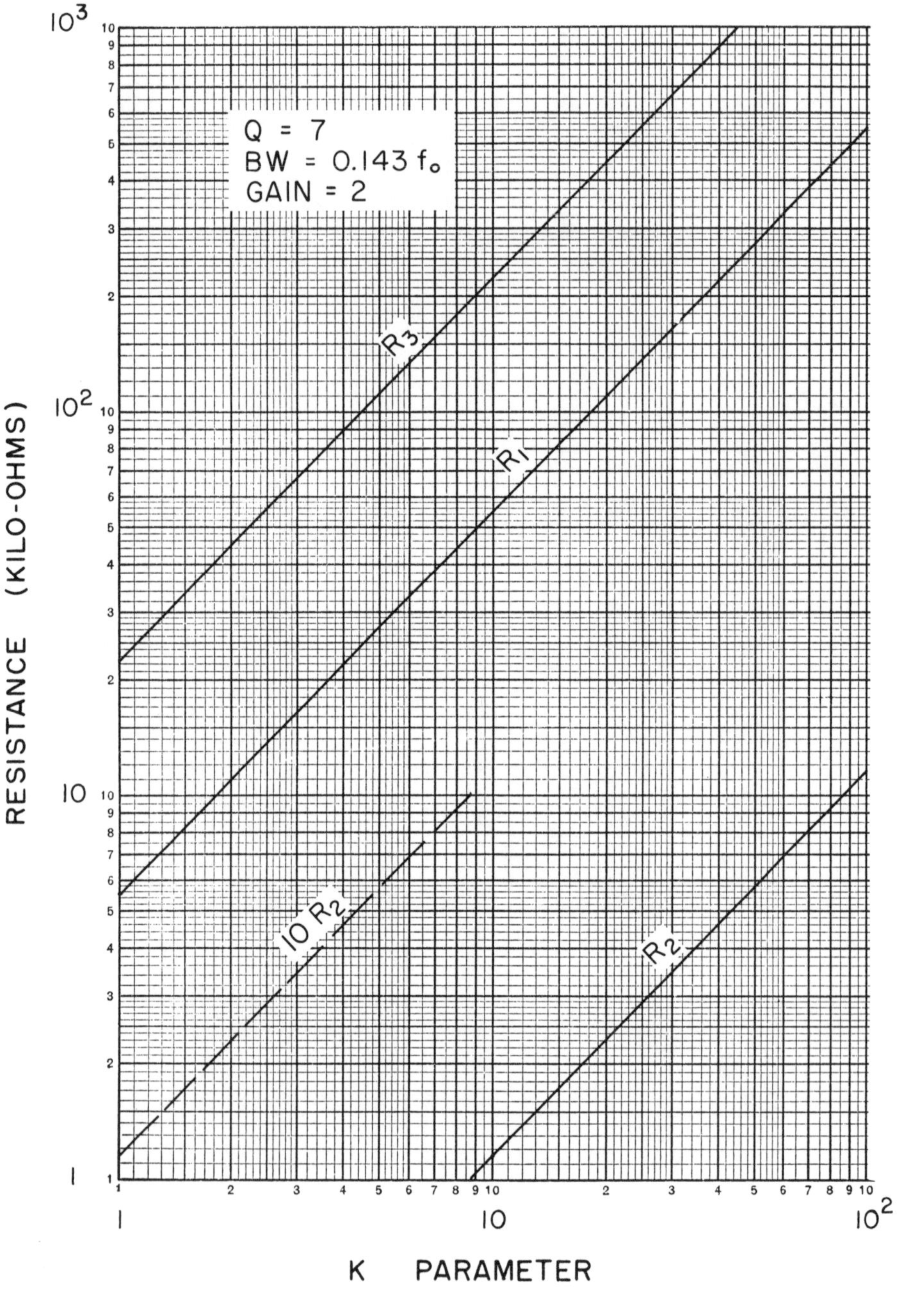
Q = 7
BW = 0.143 $f_o$
GAIN = 2
$R_3$
$R_1$
10 $R_2$
$R_2$
RESISTANCE (KILO-OHMS)
K PARAMETER
$10^3$
$10^2$
10
1
1
10
$10^2$

Fig. 4.21 Multiple-feedback band-pass filter.

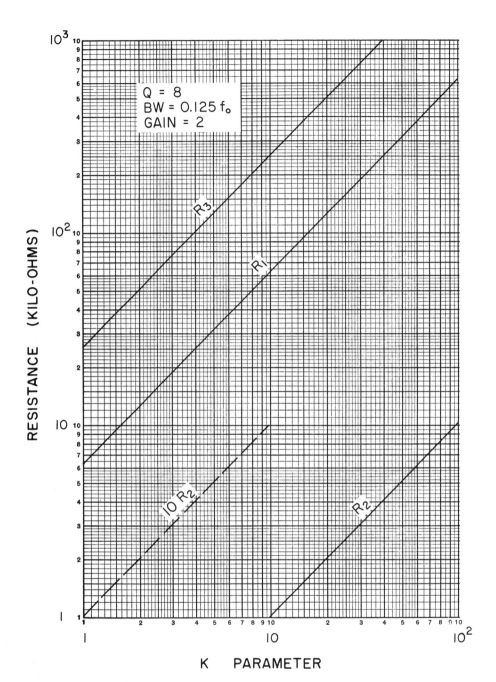

Fig. 4.22  Multiple-feedback band-pass filter.

153

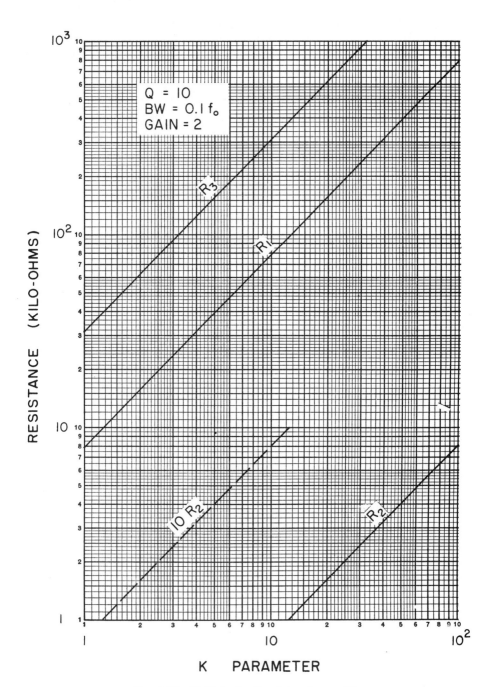

**Fig. 4.23  Multiple-feedback band-pass filter.**

154

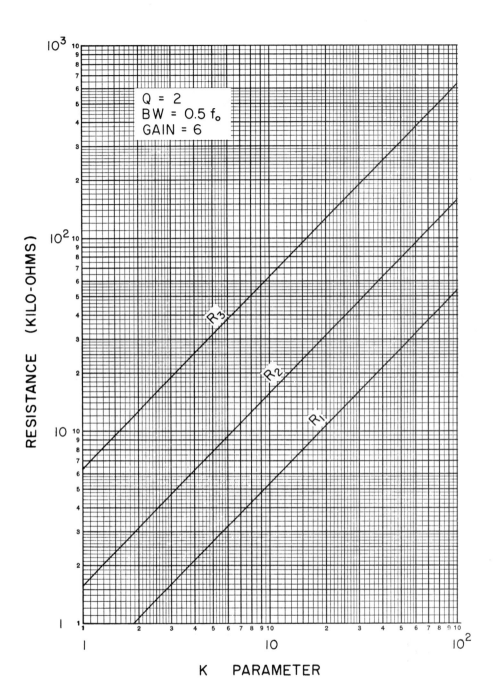

Fig. 4.24 Multiple-feedback band-pass filter.

155

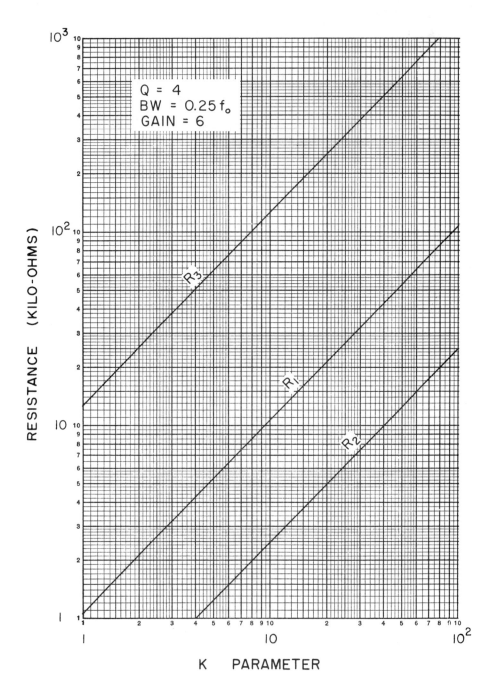

Fig. 4.25   Multiple-feedback band-pass filter.

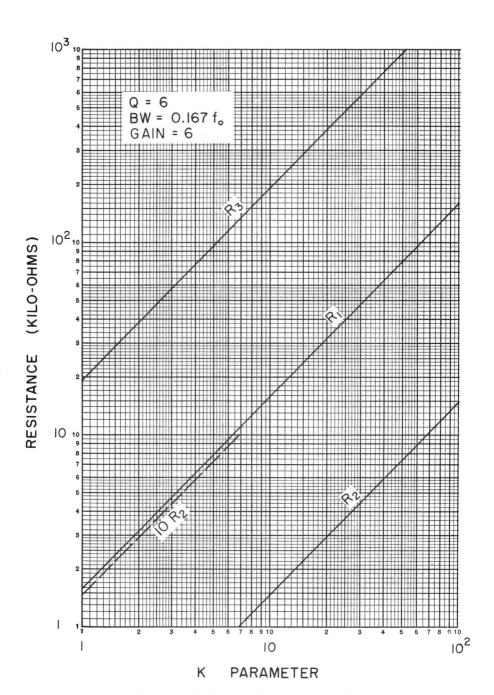

Fig. 4.26 Multiple-feedback band-pass filter.

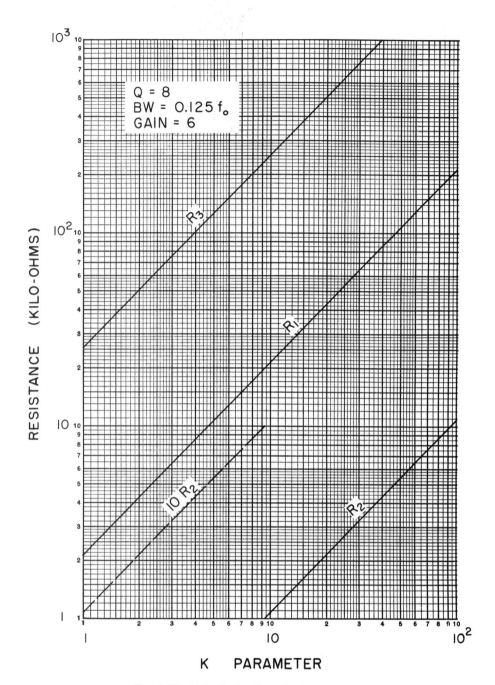

**Fig. 4.27  Multiple-feedback band-pass filter.**

K   PARAMETER

RESISTANCE  (KILO-OHMS)

Q = 8
BW = 0.125 f$_o$
GAIN = 6

R$_3$

R$_1$

10 R$_2$

R$_2$

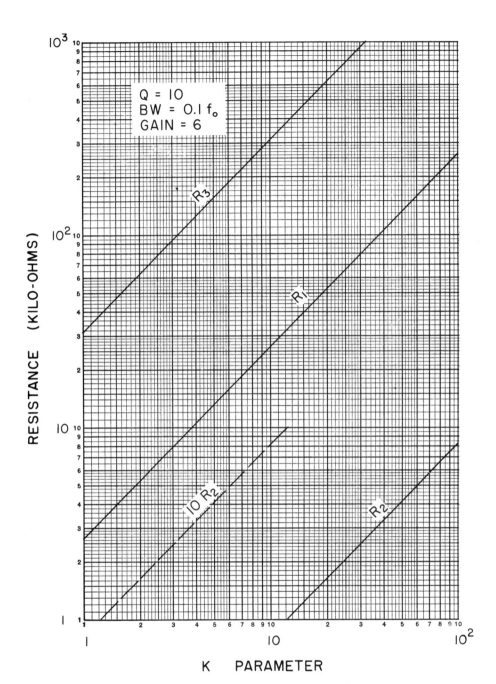

**Fig. 4.28  Multiple-feedback band-pass filter.**

159

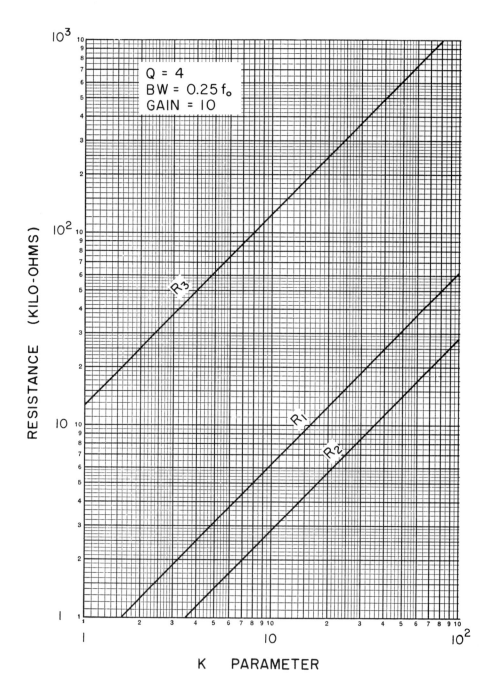

Q = 4
BW = 0.25 f$_o$
GAIN = 10

RESISTANCE (KILO-OHMS)

K   PARAMETER

**Fig. 4.29  Multiple-feedback band-pass filter.**

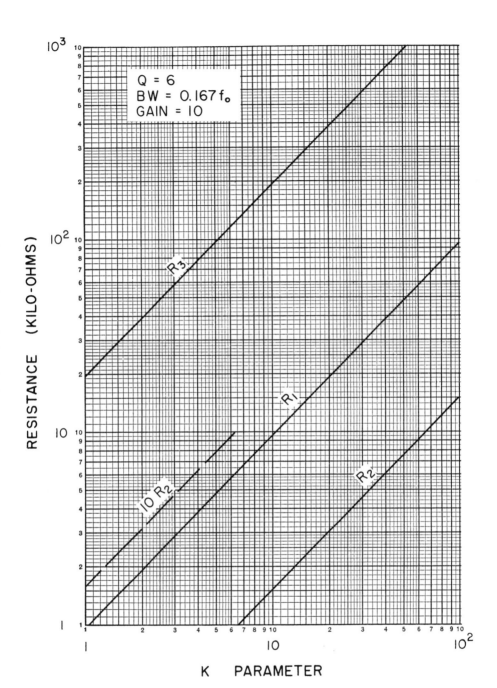

Fig. 4.30 Multiple-feedback band-pass filter.

161

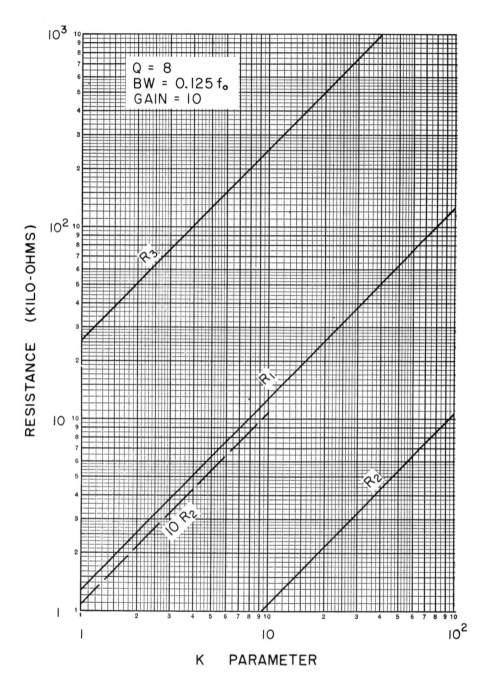

Fig. 4.31  Multiple-feedback band-pass filter.

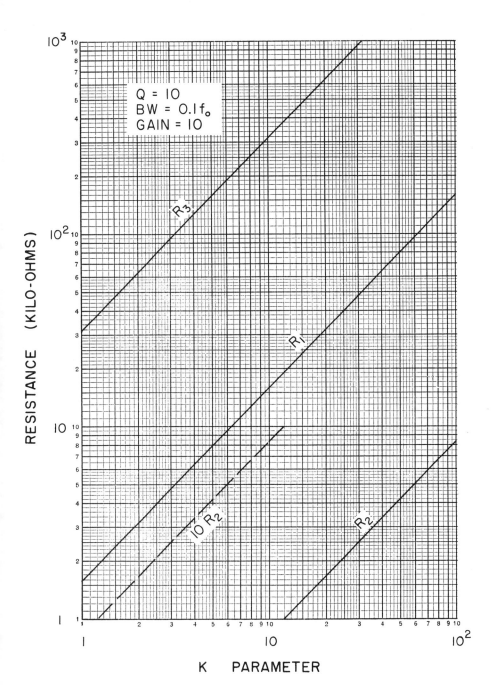

Fig. 4.32  Multiple-feedback band-pass filter.

163

## SUMMARY OF SECOND-ORDER POSITIVE-FEEDBACK BAND-PASS FILTER DESIGN PROCEDURE ( $Q \le 40$ )

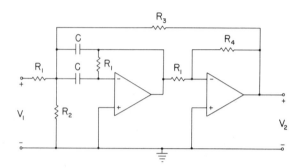

**General circuit.**

## Procedure

Given $f_0$ (Hz), $Q$ (or bandwidth $BW$ in Hz), and gain, perform the following steps:

1. Select a value of capacitance $C$ and determine a $K$ parameter from Fig. 4.12a, b, or c, as described for the second-order VCVS band-pass filter.
2. Using this value of $K$, find the resistances from the appropriate one of Figs. 4.33 to 4.40, depending on $Q$ (or $BW$) and the gain.
3. Select standard resistances which are as close as possible to those indicated on the graph and construct the circuit.

## Comments and Suggestions

The remarks given for the second-order VCVS low-pass filter are applicable with the following exceptions:

(1) The statement concerning the ratio $R_4/R_3$ is not applicable.
(2) The dc returns to ground are already satisfied by $R_1$ and $R_4$.
(3) Remarks concerning $f_c$ now apply to $f_0$.
(4) The open-loop gain of the op-amps should be at least 50 times the square root of the filter gain.

The quality factor $Q$, and hence the bandwidth $BW$, can be varied to some degree, without appreciably changing $f_0$, by varying $R_3$ or $R_4$. Also, changing $C$ slightly changes $f_0$. (See Sec. 4.4.) For minimum dc offset, one may place in the non-inverting input leads of the op-amps resistances equal to $R_1$ and $R_1 R_4 / (R_1 + R_4)$ respectively.

A specific example is given in Sec. 4.4.

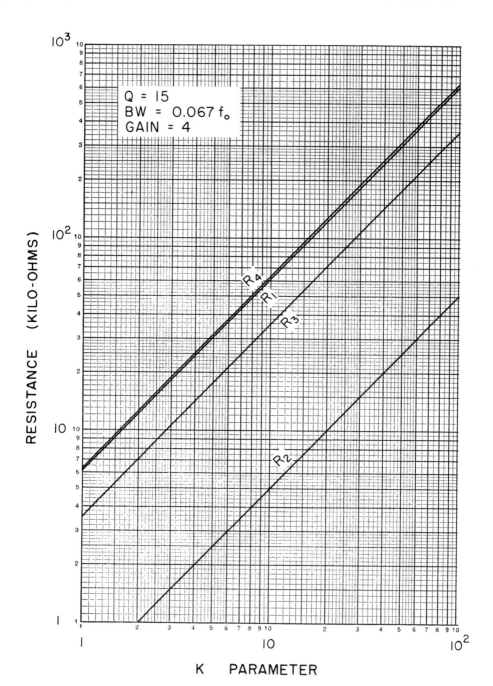

Fig. 4.33 Positive-feedback band-pass filter.

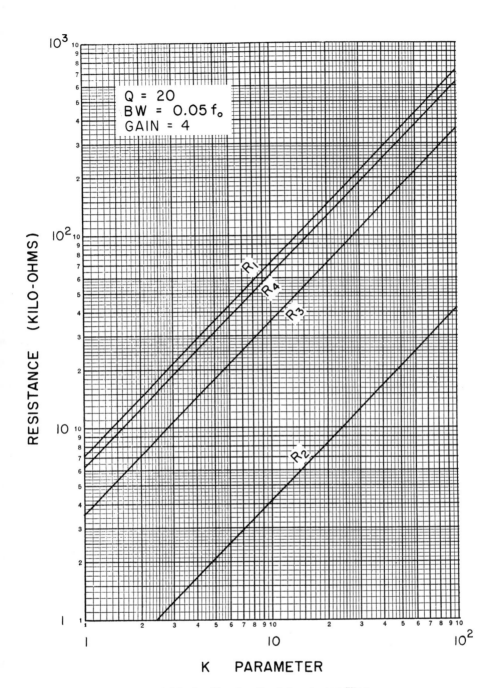

RESISTANCE (KILO-OHMS)

K     PARAMETER

Fig. 4.34   Positive-feedback band-pass filter.

167

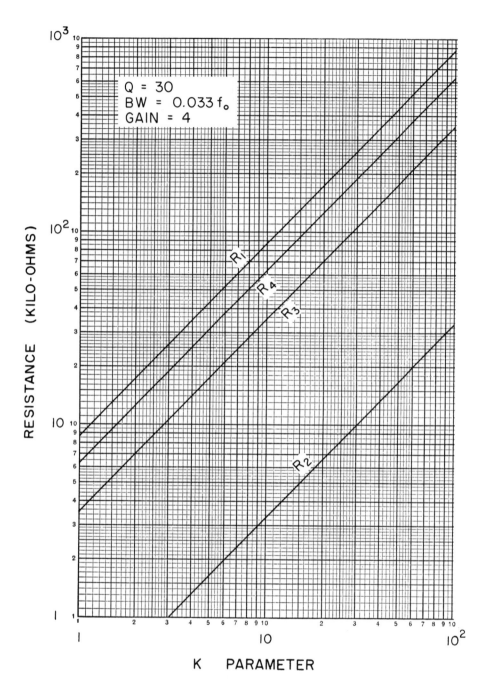

Fig. 4.35  Positive-feedback band-pass filter.

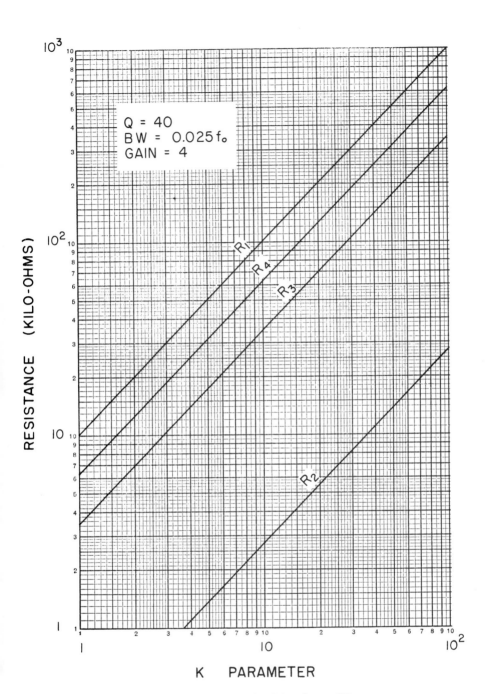

Fig. 4.36 Positive-feedback band-pass filter.

169

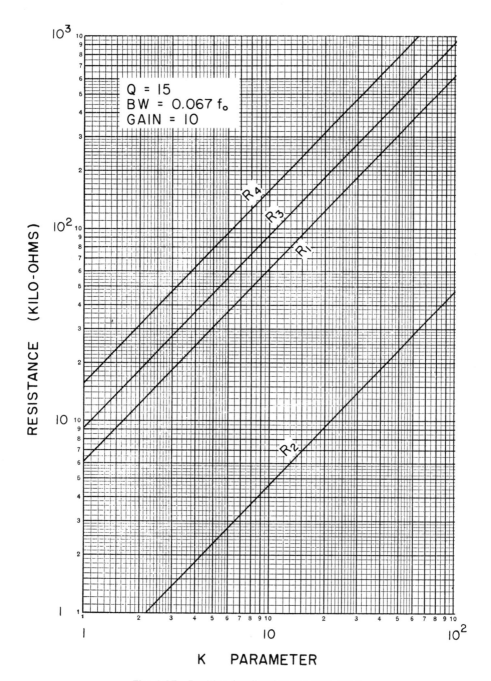

Fig. 4.37  Positive-feedback band-pass filter.

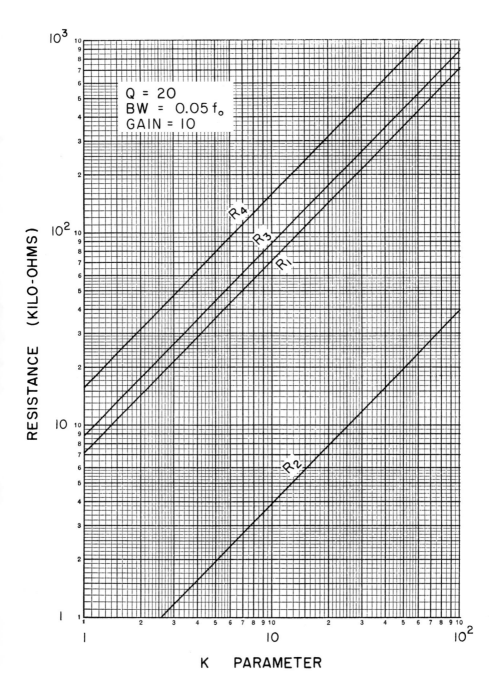

RESISTANCE (KILO-OHMS)

Q = 20
BW = 0.05 f$_o$
GAIN = 10

$R_4$

$R_3$

$R_1$

$R_2$

K  PARAMETER

Fig. 4.38  Positive-feedback band-pass filter.

171

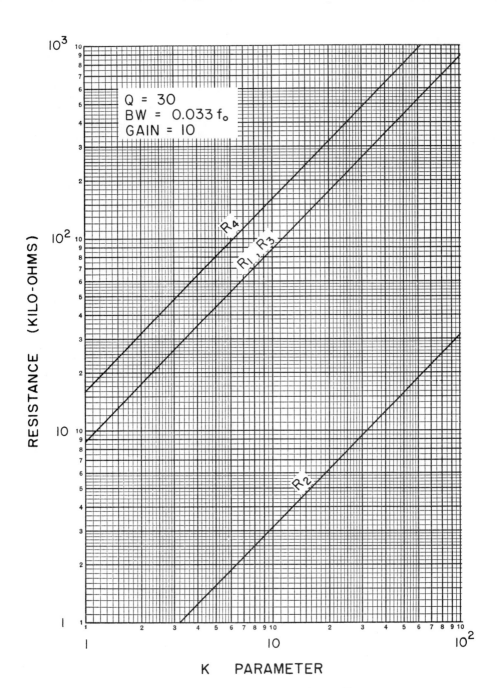

Fig. 4.39 Positive-feedback band-pass filter.

172

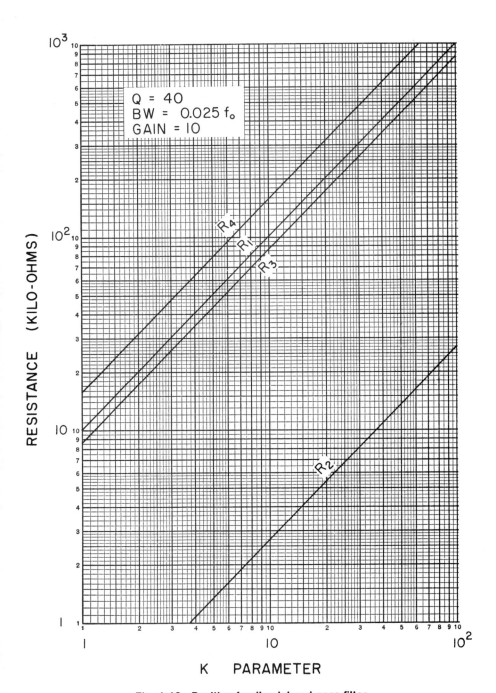

Fig. 4.40   Positive-feedback band-pass filter.

173

## SUMMARY OF SECOND-ORDER BIQUAD BAND-PASS FILTER DESIGN PROCEDURE ( $Q \leq 100$ )

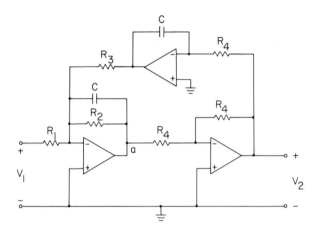

**General circuit.**

### Procedure

Given $f_0$ (Hz), $Q$ (or bandwidth $BW$ in Hz), and gain, perform the following steps:

1. Select a value of capacitance $C$ and determine a $K$ parameter from Fig. 4.12 $a$, $b$, or $c$, as described for the second-order VCVS band-pass filter.
2. Using this value of $K$ and the values of $Q$ $(f_0/BW)$ and the gain, find the resistances from Fig. 4.41.
3. Select standard resistances which are as close as possible to those indicated on the graph and construct the circuit.

### Comments and Suggestions

The remarks given for the second-order VCVS low-pass filter are applicable, except that there are three op-amps rather than one, $f_c$ should be replaced by $f_0$, and there is no resistance ratio for use in minimizing the dc offsets. Also, the dc return to ground requirement is satisfied by resistors $R_2$ and $R_3$. Finally, in the remark relative to the slew rate, $f_c$ should be replaced by $f_a$, the highest frequency of the passband.

The gain is $R_2/R_1$. If an inverting gain of the same magnitude is desired, the output may be taken at node $a$.

The filter response is readily adjusted by varying $R_1$, $R_2$, and $R_3$. Varying $R_1$ affects the gain, varying $R_2$ affects $Q$, and varying $R_3$ affects $f_0$.

The biquad band-pass filter is discussed in Sec. 4.5.

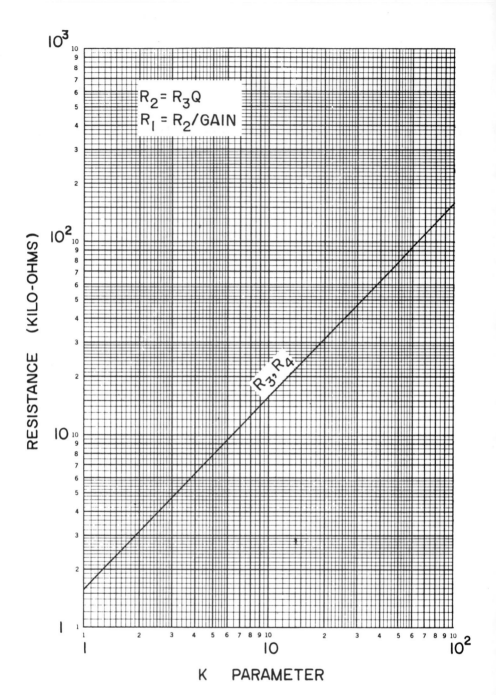

**Figure 4.41  Biquad band-pass filter.**

## SUMMARY OF HIGHER ORDER BAND-PASS FILTER DESIGN PROCEDURE

(a) $Q \leq \sqrt{2} = 1.414$

This may be accomplished by cascading a second-order low-pass filter with cutoff $f_{c2}$ and a second-order high-pass filter with cutoff $f_{c1} < f_{c2}$. For best results the frequency $f_{c2}$ should be at least twice $f_{c1}$. The procedure is described in the summaries for the second-order low- and high-pass filters.

This results in a center frequency of approximately $f_0 = \sqrt{f_{c1}f_{c2}}$ (it is exactly this if the two are both Butterworth filters), with a gain slightly less than the product of the gains of the two stages. The bandwidth is approximately $f_{c2} - f_{c1}$. (For higher $Q$, one could use fourth-order stages.)

(b) $Q \geq \sqrt{2} = 1.414$

This may be done by cascading two or more second-order band-pass filters as described in the summary sheets in this chapter.

The result is a higher $Q$ (more narrow $BW$) as given by Fig. 4.10, and a gain approximately equal to the product of the section gains. The center frequency should be that of a single stage.

# 5
## BAND-REJECT FILTERS

### 5.1 General Circuit and Equations

A band-reject filter (also called band-elimination, or notch, filter) is one which passes all frequencies except a single band. The amplitude response of such a filter is shown in Fig. 5.1, where the ideal response is that represented by the broken line, and a realizable approximation to the ideal is that represented by the solid line. The band of frequencies which is rejected is centered approximately at $\omega_0$ and its width is $B$. The bandwidth $B$ may be measured in Hz, in which case the center frequency is $f_0 = \omega_0/2\pi$ Hz. As in the band-pass case, we also define the quantity $Q$ by $\omega_0/B$ (or $f_0/B$ if $B$ is in Hz). Thus a large $Q$ indicates a small band rejected, and a small $Q$ indicates a wide band.

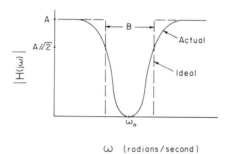

**Figure 5.1   A band-reject amplitude response.**

A second-order approximation to an ideal band-reject filter is achieved by the transfer function [21]

$$H(s) = \frac{V_2(s)}{V_1(s)} = \frac{K(s^2 + \omega_0^2)}{s^2 + Bs + \omega_0^2} \tag{5.1}$$

**179**

where $\omega_0$ is the center frequency in rad/sec and $B = \omega_0/Q$ is the width of the band rejected. The gain is defined as the value of $H(s)$ at either zero or infinity and is seen to be $K$.

A circuit which realizes Eq. (5.1) is the band-reject circuit shown in Fig. 5.2 [28], an analysis of which yields, if $R_3R_4 = 2R_1R_5$,

$$B = \frac{2}{R_4C}$$

$$\omega_0^2 = \frac{1}{R_4C^2}\left(\frac{1}{R_1} + \frac{1}{R_2}\right)$$

(5.2)

and an inverting gain of magnitude $R_6/R_3$.

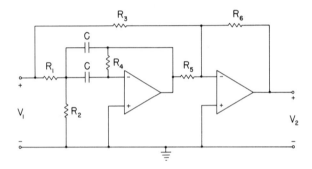

**Fig. 5.2   A band-reject filter.**

We may obtain a practical realization of the band-reject filter of Fig. 5.2, for given values of center frequency $f_0$, $Q$, and gain as described in the summary.

As an example, suppose we want $f_0 = 60$ Hz, $Q = 10$, with a gain of 10. From Fig. 5.4a, we see that if we choose a capacitance value of $C = 0.1 \ \mu\text{F}$, then the $K$ parameter is 16.6. Using this value of $K$, we find from Fig. 5.12, for a gain of 10, that the other element values are $R_1 = 131$ kΩ, $R_2 = 1.34$ kΩ, $R_3 = 16$ kΩ, $R_4 = 525$ kΩ, $R_5 = 33$ kΩ, and $R_6 = 165$ kΩ. Using standard values of 130, 1.3, 16, 510, 33, and 160 kΩ, we obtain the filter whose amplitude response is shown in Fig. 5.3. The scale used is 10 Hz/division, starting at 10 Hz. The actual results are $f_0 = 59.3$ Hz, $Q = 9.4$ ($B = 6.3$ Hz), and a gain of 10.

A summary of the techniques for obtaining a practical band-reject filter is given, together with the appropriate graphs, following this section.

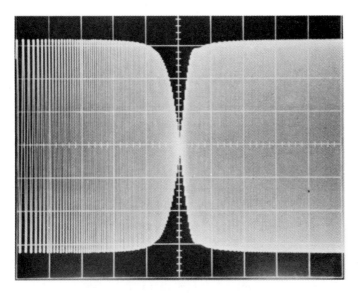

Fig. 5.3   A band-reject filter response.

## SUMMARY OF BAND-REJECT
## FILTER DESIGN PROCEDURE

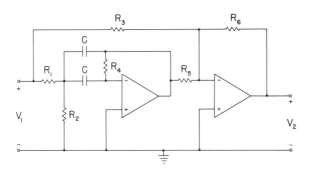

**General circuit.**

## Procedure

Given $f_0$ (Hz), $Q$ (or bandwidth $BW$), and gain, perform the following steps:

1. Select a value of capacitance $C$ and determine a $K$ parameter from Fig. 5.4a for $f_0$ between 1 and $10^2 = 100$, Fig. 5.4b for $f_0$ between 100 and $10^4 = 10,000$, and Fig. 5.4c for $f_0$ between 10,000 and $10^6 = 1,000,000$ Hz.
2. Using this value of $K$, find the resistances from the appropriate one of Figs. 5.5 through 5.13, depending on $Q$ (or $BW$) and the gain.
3. Select standard resistances which are as close as possible to those indicated on the graph and construct the circuit.

## Comments and Suggestions

The remarks given for the second-order VCVS low-pass filter are applicable with the following exceptions:

(1) The statement concerning the ratio $R_4/R_3$ is not applicable.
(2) The dc returns to ground are already satisfied by $R_4$ and $R_6$.
(3) Remarks concerning $f_c$ now apply to $f_0$.
(4) The open-loop gain of the op-amps should be at least 50 times the square root of the filter gain.

$Q$, and hence $BW$, can be varied to some degree, without appreciably changing $f_0$, by varying $R_4$. Changing $C$ slightly changes $f_0$ slightly. For minimum dc offset, one may place, in the noninverting input leads of the op-amps, resistances equal to $R_4$ and $R_5 R_6/(R_5 + R_6)$ respectively.

A specific example is given in Sec. 5.1.

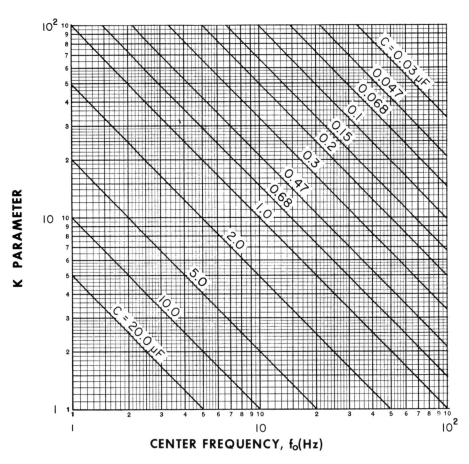

Fig. 5.4 (*a*)  *K* parameter versus frequency.

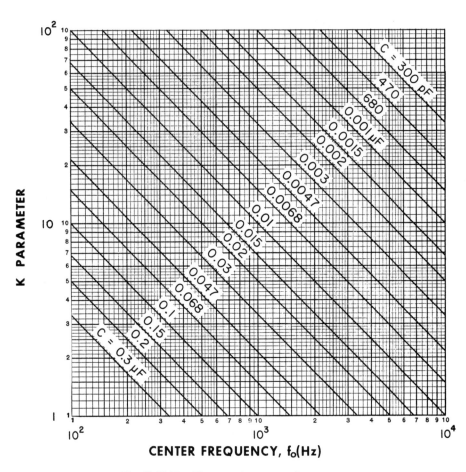

**K PARAMETER**

**CENTER FREQUENCY, f$_o$(Hz)**

**Fig. 5.4(b)** *K* parameter versus frequency.

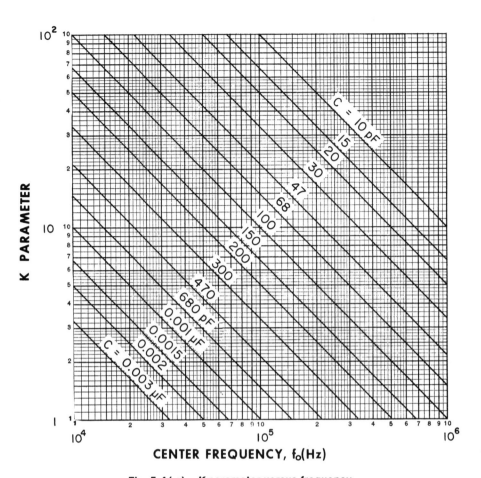

Fig. 5.4(c)   K parameter versus frequency.

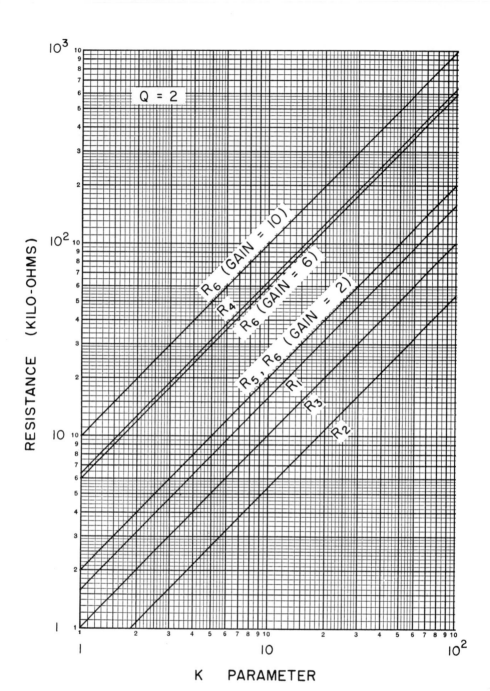

**Fig. 5.5   Band-reject filter.**

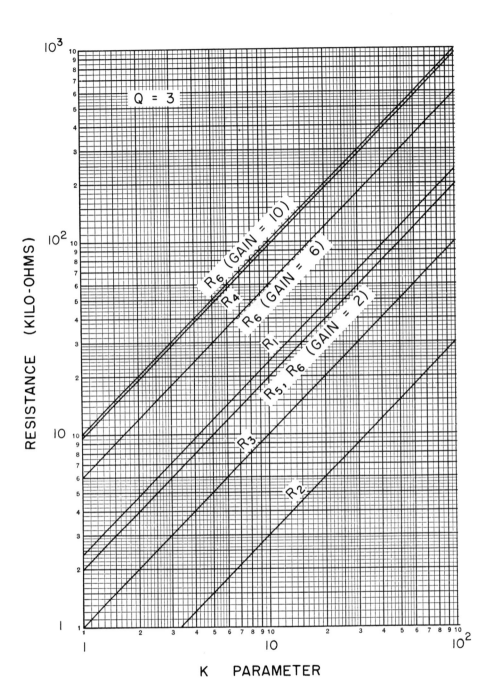

Fig. 5.6  Band-reject filter.

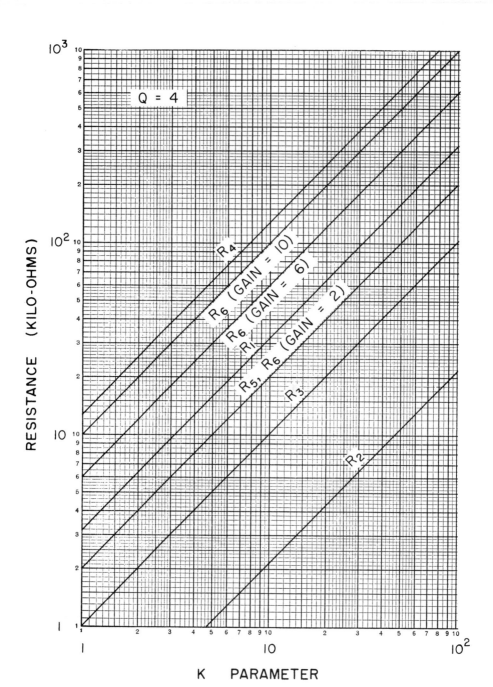

Fig. 5.7  Band-reject filter.

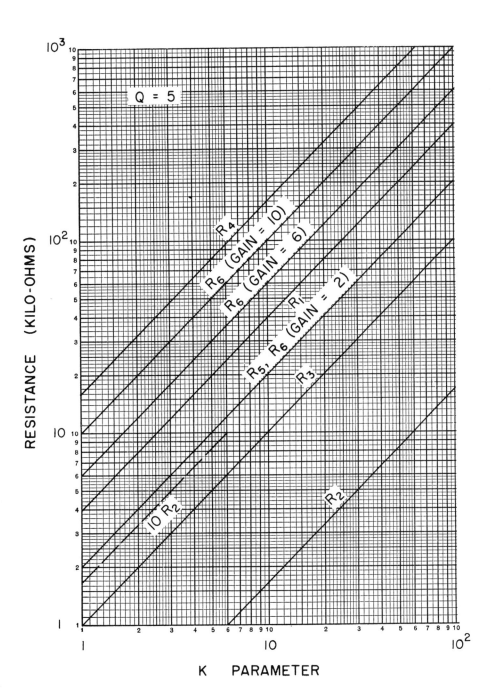

**Fig. 5.8  Band-reject filter.**

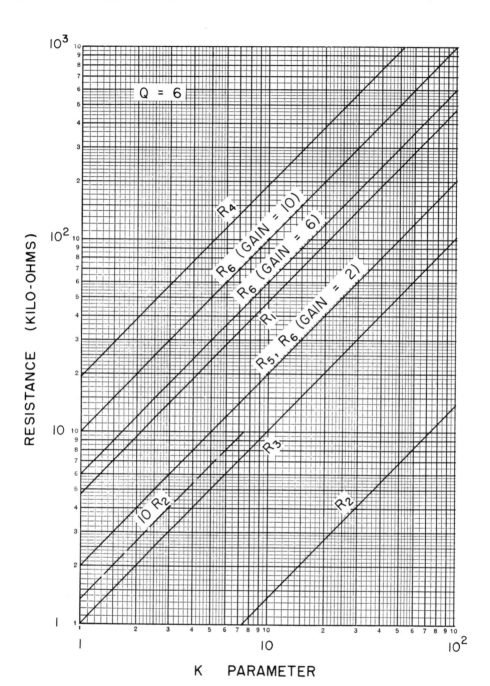

RESISTANCE (KILO-OHMS)

$Q = 6$

$R_4$

$R_6$ (GAIN = 10)

$R_6$ (GAIN = 6)

$R_1$

$R_5$, $R_6$ (GAIN = 2)

$R_3$

$10 R_2$

$R_2$

$10^3$

$10^2$

$10$

$1$

$1$      $10$      $10^2$

K    PARAMETER

Fig. 5.9   Band-reject filter.

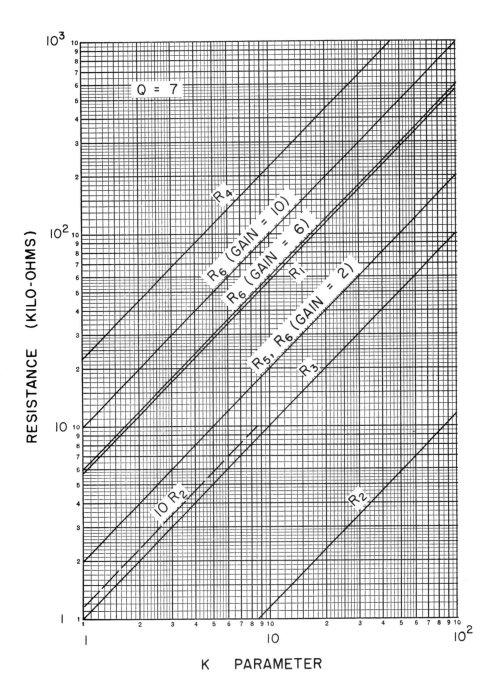

**Fig. 5.10 Band-reject filter.**

191

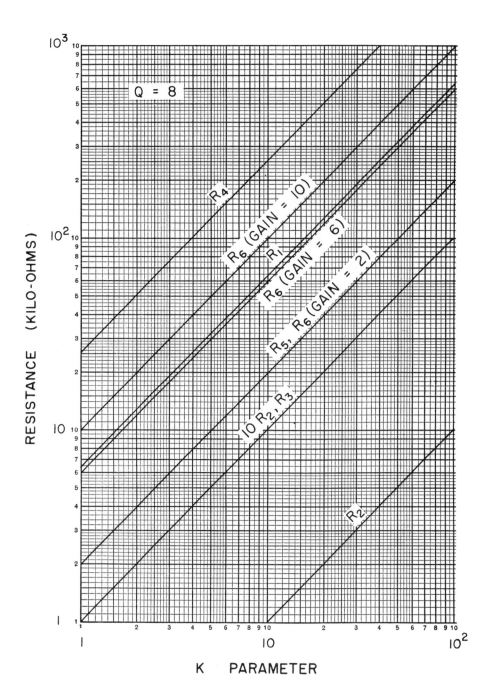

**Fig. 5.11 Band-reject filter.**

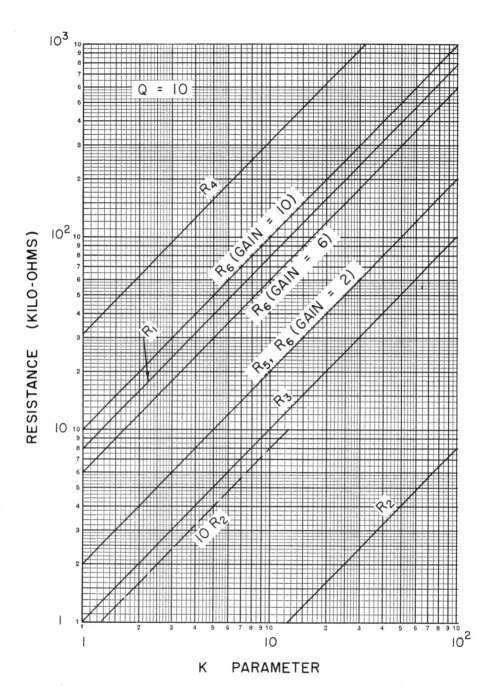

Fig. 5.12 Band-reject filter.

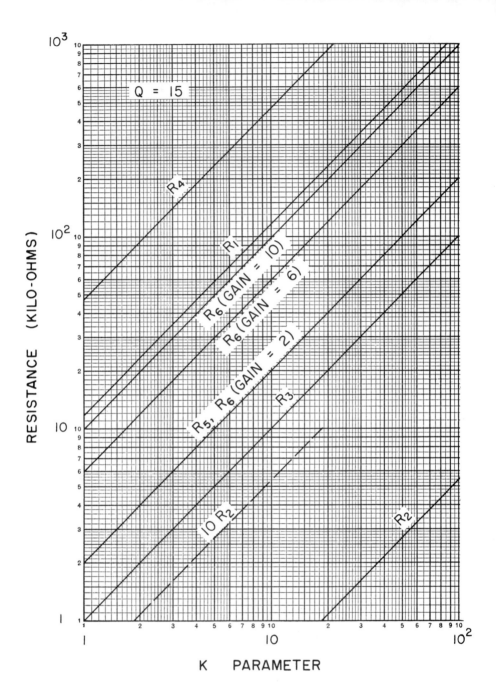

**Fig. 5.13 Band-reject filter.**

# 6
## PHASE-SHIFT AND TIME-DELAY FILTERS

### 6.1 All-Pass (Phase-Shift) Filters

An all-pass, or phase-shifting, filter is one which passes signals of all frequencies equally well while changing or shifting their phase by some prescribed amount. Since shifting a frequency by some negative amount is equivalent to delaying that component by some positive time as it passes through the filter, the all-pass filter may also be thought of as a time-delay circuit. The phase shift or time delay of its transfer function varies with frequency as the amplitude remains essentially fixed over the useful range of frequencies.

Our transfer functions are ratios of output to input voltages, $V_2/V_1$. At $\omega_0$ (or in Hz, $f_0$), if the phase shift is a negative number, say $\phi(\omega_0) = -\phi_0$ degrees, then at $\omega_0$ the phase of the input $V_1$ is greater than that of the output $V_2$ by $\phi_0$ degrees. Thus, if the two waveforms are viewed simultaneously, the input wave reaches its peaks or dips $\phi_0$ degrees before the output wave reaches its peaks or dips. Therefore the input signal is leading the output signal by $\phi_0$ degrees. Also the difference in time in seconds between a peak or dip of the input wave and the immediately preceding peak or dip of the output wave, when the amplitudes of the two waves are plotted versus time, is the time delay. Evidently a phase shift of $-\phi_0$ is equivalent to a phase shift of $360° - \phi_0$. For example, if the input wave leads the output by $270°$ ($\phi = -\phi_0 = -270°$), then it may also be said that the input leads by $-90°$ ($\phi = -\phi_0 = +90°$), in which case the output leads by $+90°$.

The amplitude response of the transfer function of an all-pass filter is ideally constant for all frequencies and in a practical realization should be very nearly constant over the range of operation. The phase response might typically look like that of Fig. 6.1, which is plotted for $0 \geq \phi \geq -360°$. One should remember that these values are equivalent to values obtained by adding to them multiples of 360°. Figure 6.2 shows an output wave lagging an input wave (or the input is leading the output) by an amount $\phi_0$. If the horizontal axis were in time (seconds), the difference between the two successive peaks would be time delay.

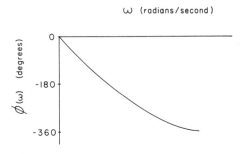

Fig. 6.1   A typical phase response.

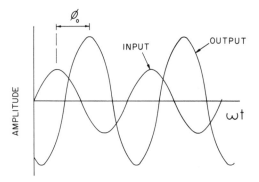

Fig. 6.2   Two waves with different phases.

A second-order approximation to an all-pass filter is achieved by the transfer function [21]

$$H(s) = \frac{V_2(s)}{V_1(s)} = \frac{K(s^2 - as + b)}{s^2 + as + b} \tag{6.1}$$

where $a$ and $b$ are appropriately chosen constants. The phase shift $\phi(\omega)$, which could also be given in terms of $f = \omega/2\pi$ Hz, is given by

$$\phi(\omega) = -2 \arctan\left(\frac{a\omega}{b - \omega^2}\right) \tag{6.2}$$

The amplitude is $|H(j\omega)| = K$, which is also the gain of the filter.

An all-pass network which has a transfer function such as Eq. (6.1) for appropriate values of the circuit elements is shown in Fig. 6.3 [29]. Analysis of the circuit yields

$$a = \frac{2}{R_2 C} \tag{6.3}$$

$$b = \frac{1}{R_1 R_2 C^2}$$

and a gain of $K = R_4/(R_3 + R_4)$, provided that $R_2 R_3 = 4R_1 R_4$.

The designer of an all-pass circuit such as that of Fig. 6.3 may specify the phase shift desired at a given frequency and obtain practical element values as described in the summary at the end of the chapter. We have constructed the graphs to yield a gain of ½ in every case.

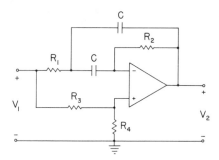

**Fig. 6.3   An all-pass filter.**

As an example, suppose we wish to construct an all-pass filter with a phase shift of 90° (equivalent to $-270°$) at $f_0 = 1000$ Hz and we have 0.01 $\mu$F capacitors available. From Fig. 6.7$b$, corresponding to $f_0$ and a 0.01 $\mu$F capacitance, we obtain the $K$ parameter of 10. Using this value of $K$, we see from Fig. 6.16 that the resistances are $R_1 = 12.9$ k$\Omega$, $R_2 = 51.5$ k$\Omega$, and $R_3 = R_4 = 103$ k$\Omega$. Using standard values of 13, 51, and 100 k$\Omega$ respectively, with an SN72741N op-amp we obtain an actual phase shift of 86° at $f = 1000$ Hz. The useful range of operation of the circuit is from 0 to 160,000 Hz, at which point the amplitude response drops below 0.707 times its initial value. Figure 6.4 shows the input and output waves at $f = 1000$ Hz, where consecutive peaks are approximately one division apart. The scale used is 90°/division. Since the gain is ½, the output wave has half the amplitude of the input wave, and may be seen to be leading, as required.

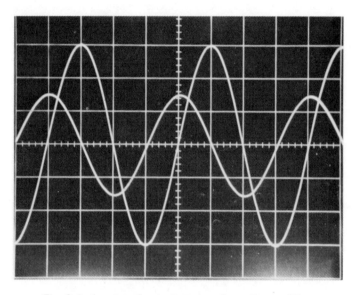

**Fig. 6.4  Input and output waves of an all-pass filter.**

### 6.2  Bessel (Constant-Time-Delay) Filters

A filter for which the phase response shown in Fig. 6.1 becomes a straight line sloping downward to the right is a linear-phase, or constant-time-delay, filter. It has the property that its phase shift is proportional to the frequency, and the time delay, described in the previous section, is constant for all frequencies.

A filter whose transfer function is a constant divided by a polynomial and which best approximates a constant-time-delay circuit is the Bessel filter [30]. In the second-order case it has a transfer function

$$H(s) = \frac{V_2(s)}{V_1(s)} = \frac{K}{s^2 + 3\omega_0 s + 3\omega_0^2} \tag{6.4}$$

and a very nearly linear phase response and constant time delay for frequencies from 0 to $f_0 = \omega_0/2\pi$ Hz. The amplitude response resembles somewhat that of an all-pass filter except that it drops slowly and monotonically from its maximum value of $K/3\omega_0^2$ at zero frequency.

To illustrate the quality of the Bessel filter, we note that if Eq. (6.4) is achieved, then the phase shift varies almost linearly from 0° at 0 Hz to −56.3° (almost 1 radian) at $f_0$. The time delay is $159.15/f_0$ milliseconds (msec) at 0 Hz, and is 99.96% of this at $f_0/4$ Hz, 99.4% at $f_0/2$ Hz, 97.1% at $3f_0/4$ Hz, and is 92.3% at $f_0$ Hz. Indeed, the time delay is almost constant up to $2f_0$.

The Bessel filter transfer function is identical in form to that of a second-order low-pass filter, considered in Chapter 2. Therefore we may realize Eq. (6.4) with the same Sallen and Key circuit, which for convenience we repeat as Fig. 6.5. Analysis of the circuit shows that it realizes Eq. (6.4) for

$$K = \frac{\mu}{R_1 R_2 C C_1}$$

$$3\omega_0 = \frac{1}{R_2 C_1}(1 - \mu) + \frac{1}{R_1 C} + \frac{1}{R_2 C} \qquad (6.5)$$

$$3\omega_0^2 = \frac{1}{R_1 R_2 C C_1}$$

where

$$\mu = 1 + \frac{R_4}{R_3} \qquad (6.6)$$

is the gain of the filter. The low-pass biquad circuit of Fig. 2.4 may also be used for the Bessel filter.

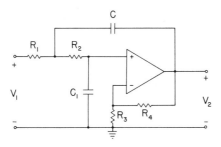

**Fig. 6.5   A second-order Bessel filter.**

A practical realization of Fig. 6.5 may be obtained for a given time delay and gain as described in the summary at the end of the chapter.

As an example, suppose we wish to obtain a filter with a constant time delay of 31.0 microseconds ($\mu$sec) and a gain of 2. From Fig. 6.17$b$, we see that $f_0 = 5000$ Hz. Selecting $C = 0.01$ $\mu$F, we have from Fig. 6.7$b$, a $K$ parameter of 2. Finally from Fig. 6.18, we have $C_1 = C = 0.01$ $\mu$F, $R_1 = 1.06$ k$\Omega$, $R_2 = 3.2$ k$\Omega$, and $R_3 = R_4 = 8.5$ k$\Omega$. Using standard values of 1.1, 3.3, and 8.2 k$\Omega$ and an MC1741 op-amp, we obtain an actual time delay of 31 $\mu$sec at 5000 Hz. Both the amplitude response and the input and output waveforms are shown in Fig. 6.6. The scales are 500 Hz/division and 50 $\mu$sec/division (90°/division) respectively.

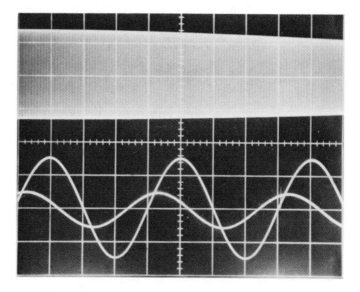

**Fig. 6.6    Amplitude response and waveforms of a Bessel filter.**

A summary of the procedures considered in this chapter is given, along with the graphs, following this section.

## SUMMARY OF ALL-PASS (PHASE-SHIFT) FILTER DESIGN PROCEDURE

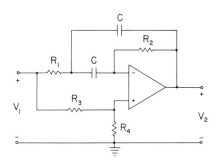

**General circuit.**

### Procedure

Given $f_0$ (Hz) and the phase shift $\phi$ desired at $f_0$, perform the following steps:

1.  Select a value of capacitance $C$ and determine a $K$ parameter from Fig. 6.7a if $f_0$ is between 1 and $10^2 = 100$, from Fig. 6.7b if $f_0$ is between 100 and $10^4 = 10,000$, and from Fig. 6.7c if $f_0$ is between 10,000 and $10^6 = 1,000,000$ Hz.
2.  Using this value of $K$, find the resistances from the appropriate one of Figs. 6.8 through 6.16, depending on $\phi$.
3.  Select standard resistances which are as close as possible to those indicated on the graph and construct the circuit.

### Comments and Suggestions

The remarks given for the second-order VCVS low-pass filter are applicable with the following exceptions:

(1)  The statement concerning the ratio $R_4/R_3$ is not applicable.
(2)  The dc return to ground is already satisfied by $R_4$.
(3)  Remarks concerning $f_c$ now apply to $f_0$.

For tuning purposes to a limited degree, varying $R_2$ with a potentiometer affects $\phi$.

A specific example is given in Sec. 6.6.

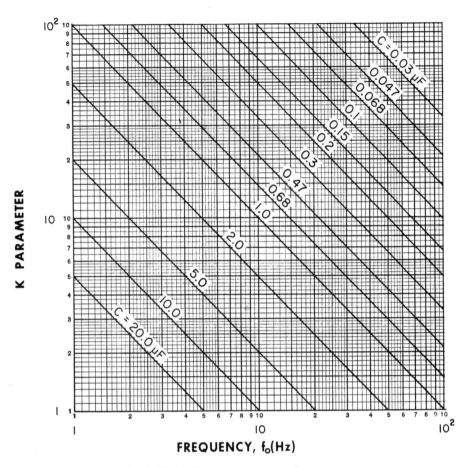

Fig. 6.7 (*a*)   *K* parameter versus frequency.

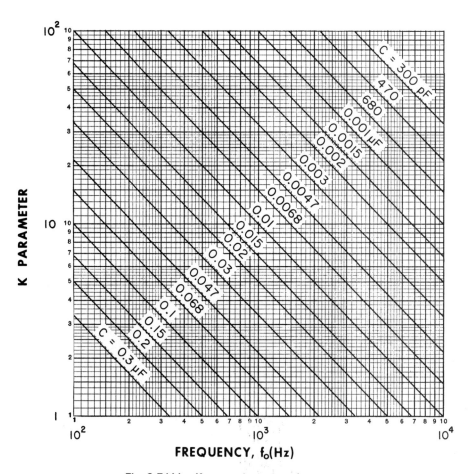

**K PARAMETER**

**FREQUENCY, f₀(Hz)**

Fig. 6.7 (*b*)  *K* parameter versus frequency.

203

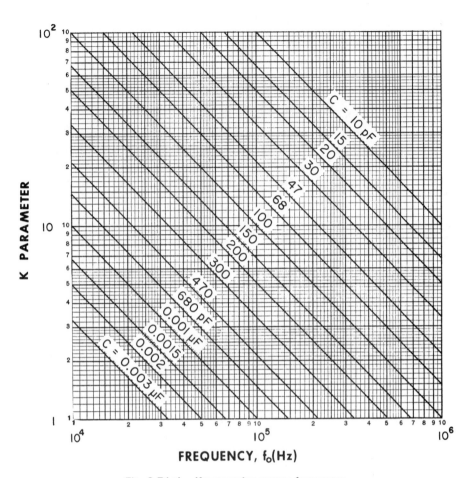

Fig. 6.7 ( c )  *K* parameter versus frequency.

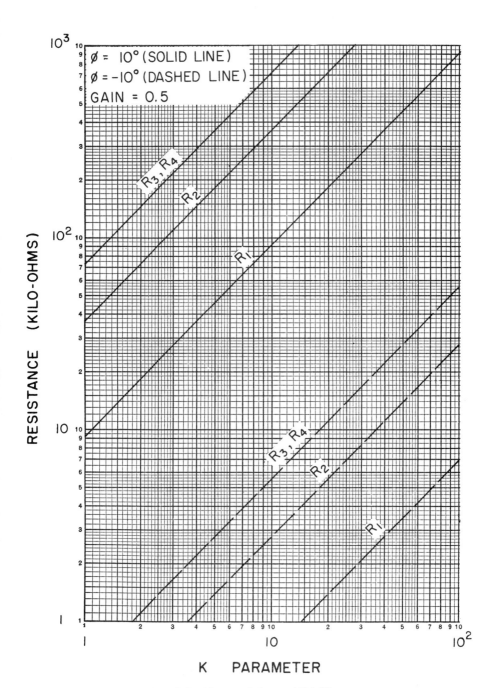

Fig. 6.8 All-pass (phase-shift) filter.

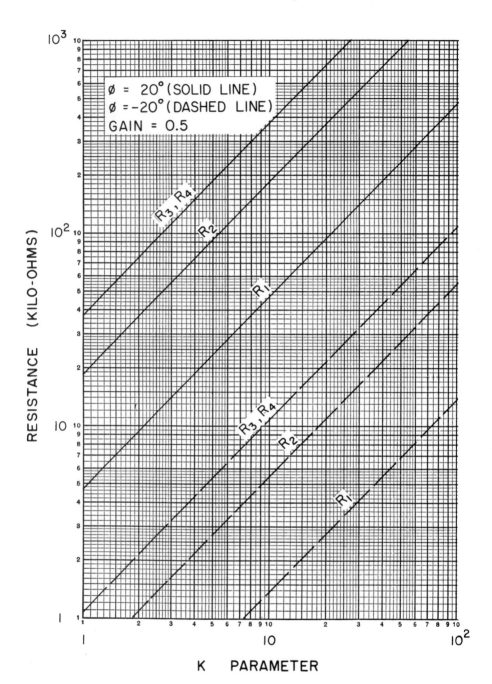

Fig. 6.9 All-pass (phase-shift) filter.

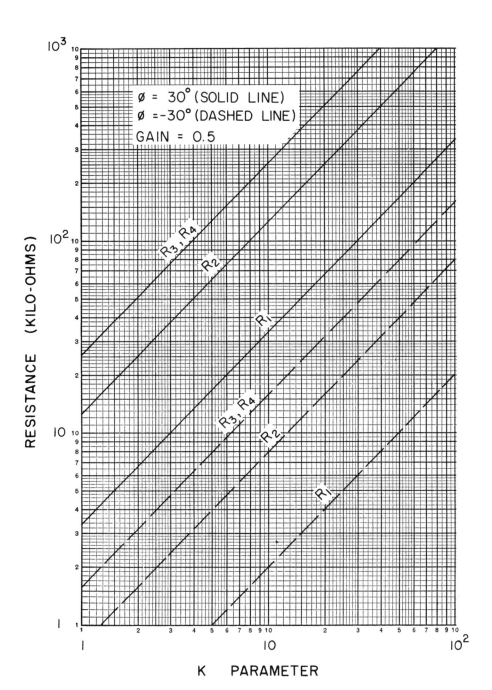

RESISTANCE (KILO-OHMS)

ø = 30° (SOLID LINE)
ø = -30° (DASHED LINE)
GAIN = 0.5

$R_3, R_4$
$R_2$
$R_1$
$R_3, R_4$
$R_2$
$R_1$

K    PARAMETER

Fig. 6.10  All-pass (phase-shift) filter.

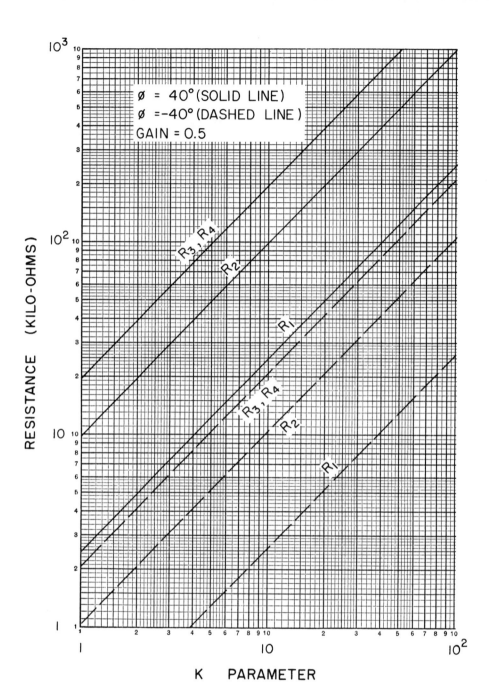

Fig. 6.11 All-pass (phase-shift) filter.

208

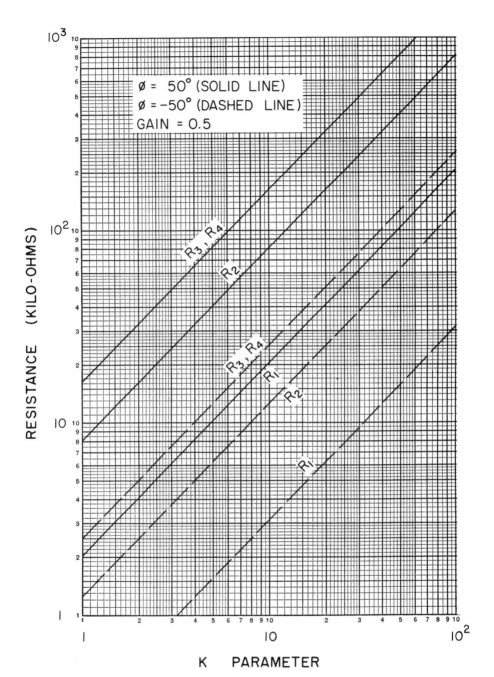

**Fig. 6.12 All-pass (phase-shift) filter.**

209

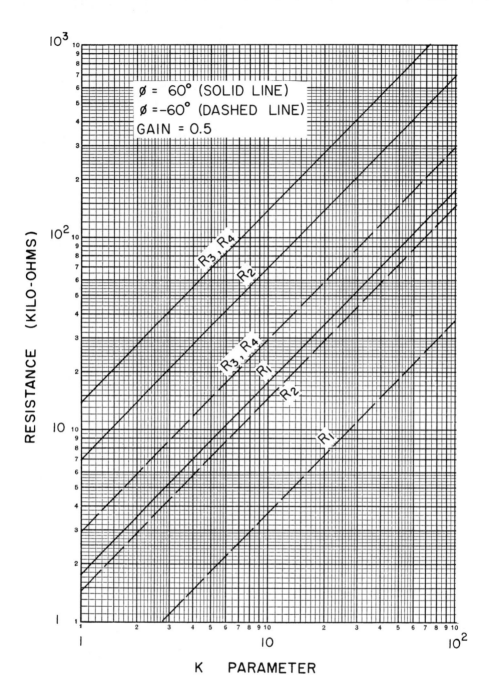

**Fig. 6.13  All-pass (phase-shift) filter.**

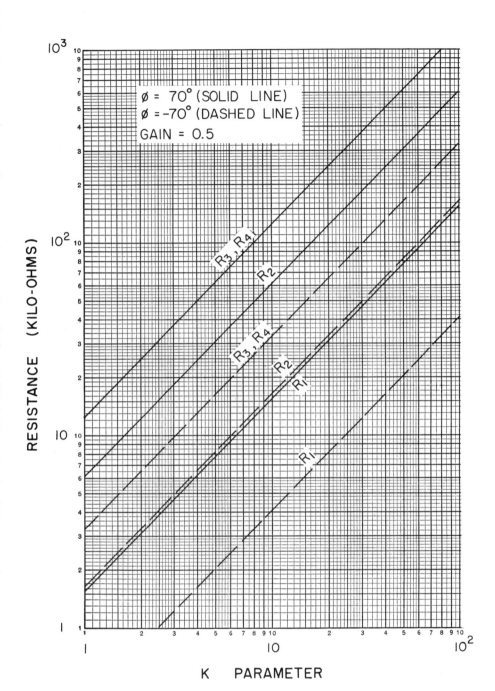

Fig. 6.14 All-pass (phase-shift) filter.

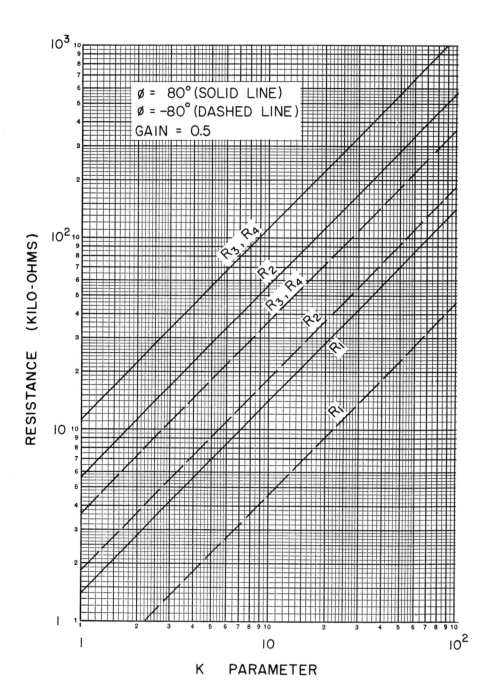

Fig. 6.15 All-pass (phase-shift) filter.

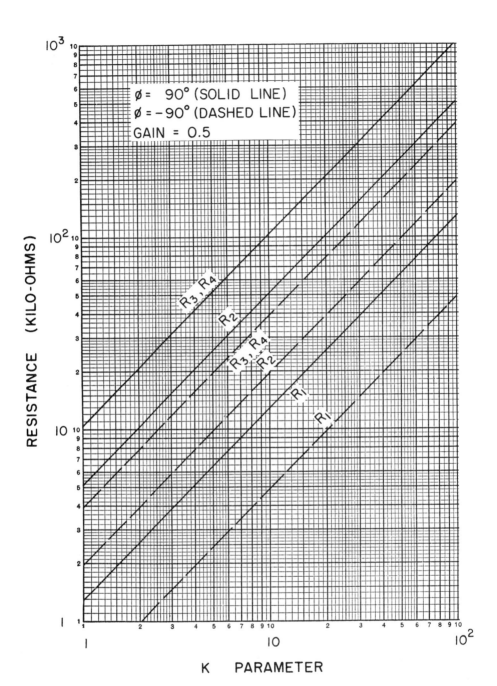

Fig. 6.16 All-pass (phase-shift) filter.

## SUMMARY OF BESSEL
## (CONSTANT-TIME-DELAY)
## FILTER DESIGN PROCEDURE

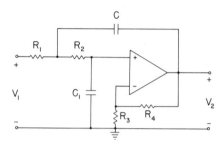

**General circuit.**

### Procedure

Given time delay $T_d$ and gain, perform the following steps:

1. Find $f_0$ from Fig. 6.17$a$ if $T_d$ is between 1.47 and 147 msec, from Fig. 6.17$b$ if $T_d$ is between 14.7 $\mu$sec and 1.47 msec, and from Fig. 6.17$c$ if $T_d$ is between 0.147 and 14.7 $\mu$sec. (Note: Alternately, one could start with $f_0$ and find $T_d$ by this procedure.)

2. Select a value of capacitance $C$ and determine a $K$ parameter from the appropriate one of Figs. 6.7$a$, $b$, or $c$, as described in the all-pass filter summary.

3. Using this value of $K$, find the capacitance $C_1$ and the resistance values from the appropriate one of Figs. 6.18, 6.19, or 6.20, depending on the gain.

4. Select standard resistances which are as close as possible to those indicated on the graph and construct the circuit.

### Comments and Suggestions

The remarks given for the second-order VCVS low-pass filter are applicable with the exception that those concerning $f_c$ now apply to $f_0$.

For tuning purposes to a limited degree, varying $R_2$ with a potentiometer affects $T_d$.

A specific example is given in Sec. 6.2.

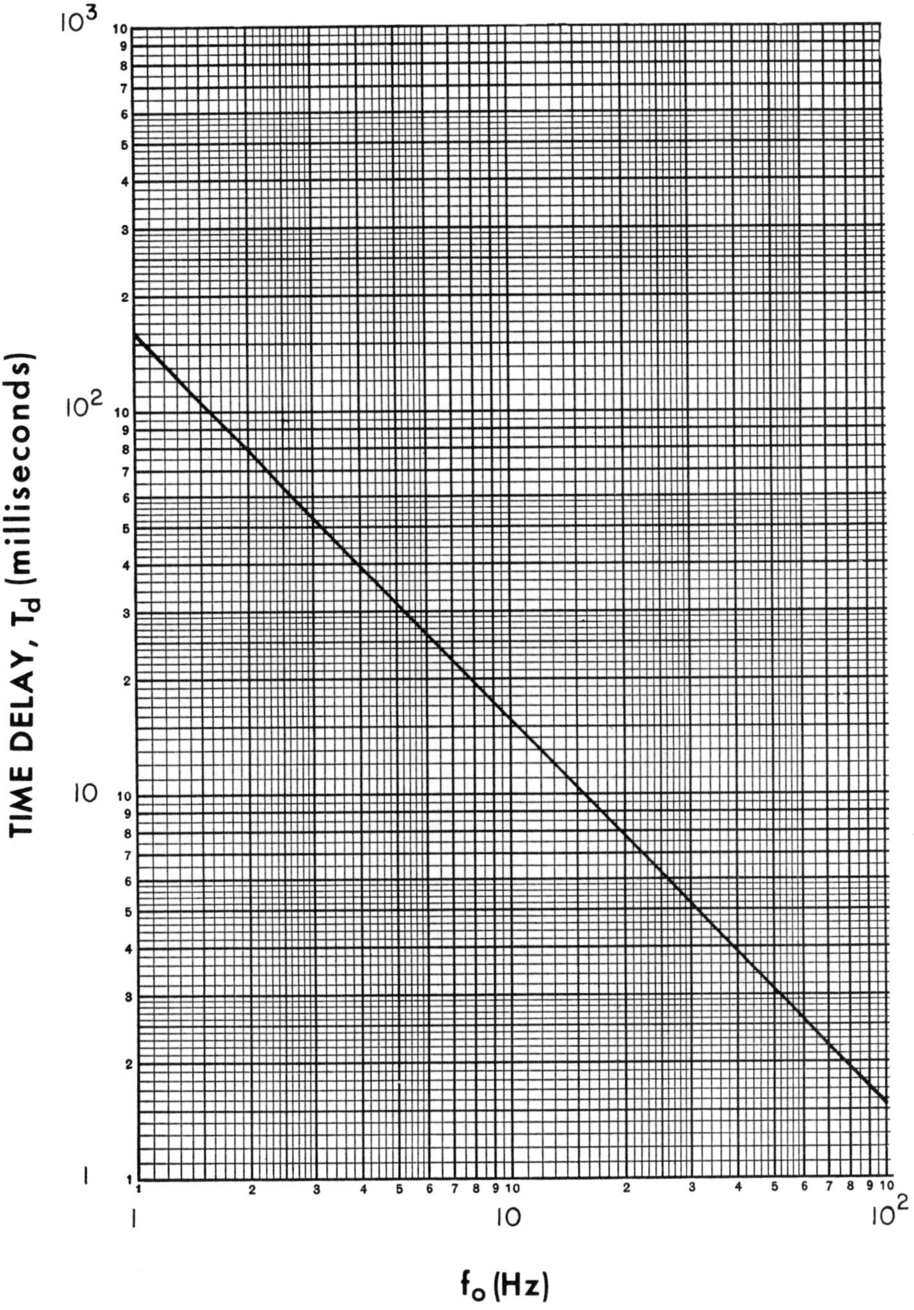

**Fig. 6.17 (a)   Time delay versus frequency.**

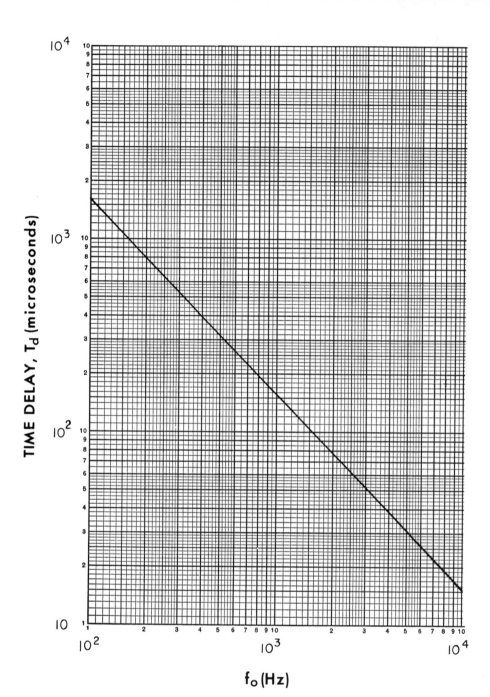

**Fig. 6.17 (b)    Time delay versus frequency.**

216

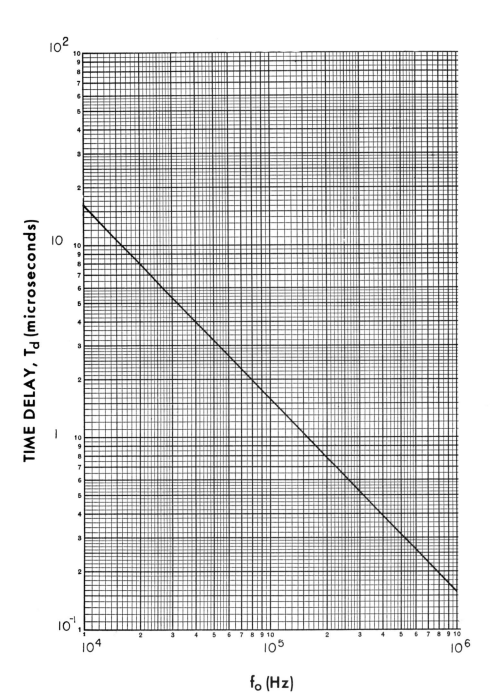

**Fig. 6.17 ( c )   Time delay versus frequency.**

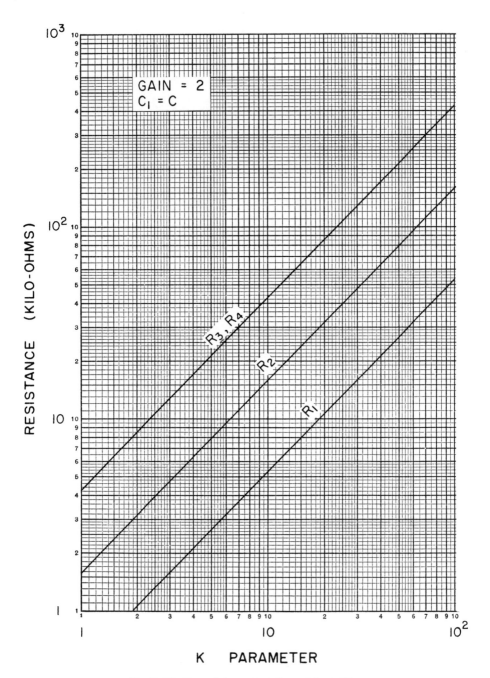

GAIN = 2
$C_1 = C$

$R_3, R_4$

$R_2$

$R_1$

RESISTANCE (KILO-OHMS)

K    PARAMETER

Fig. 6.18   Bessel (constant-time-delay) filter.

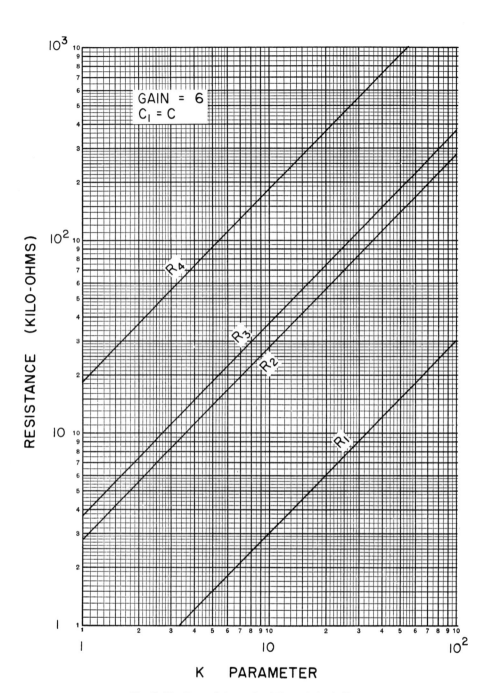

Fig. 6.19 Bessel (constant-time-delay) filter.

219

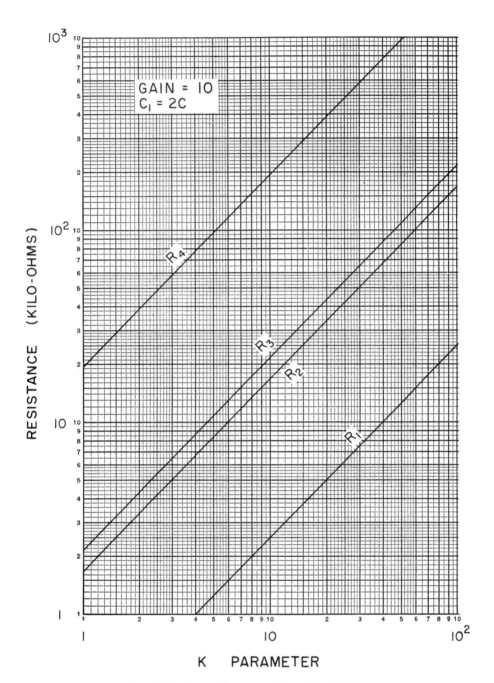

**Fig. 6.20 Bessel (constant-time-delay) filter.**

# BIBLIOGRAPHY

1. *Linear Integrated Circuit D.A.T.A. Book.* Derivation and Tabulation Associates, Inc., Orange, N.J., 1981.
2. G. E. Tobey, J. G. Graeme, and L. P. Huelsman (eds.), *Operational Amplifiers: Design and Applications*, McGraw-Hill Book Co., New York, 1971.
3. *RCA Linear Integrated Circuits*, RCA Solid State Division, Somerville, N.J. (current edition).
4. R. Melen and H. Garland, *Understanding IC Operational Amplifiers* (2d ed.), Howard W. Sams and Co., New York, 1978.
5. *Radio Electronics* (monthly periodical), Gernsback Publications, Inc., New York.
6. S. K. Mitra, *Analysis and Synthesis of Linear Active Networks*, John Wiley and Sons, Inc., New York, 1969.
7. F. C. Fitchen, *Electronic Integrated Circuits and Systems*, Van Nostrand Reinhold Co., New York, 1970.
8. A. B. Grebene, *Analog Integrated Circuit Design*, Van Nostrand Reinhold Co., New York, 1972 (reprinted by Robert E. Krieger Publishing Co., Melbourne, Fla., 1978).
9. *Popular Electronics* (monthly periodical), Ziff-Davis Publishing Co., New York.
10. *Electronics* (biweekly periodical), McGraw-Hill Book Co., New York.
11. G. S. Moschytz and R. W. Wyndrum Jr., "Applying the operational amplifier," *Electronics*, December 9, 1968, pp. 98–106.
12. D. E. Johnson and V. Jayakumar, *Operational Amplifier Circuits: Design and Application*, Prentice-Hall, Inc., Englewood Cliffs, N.J., 1982.
13. *Motorola Linear Integrated Circuits*, Motorola Semiconductor Products, Inc., Phoenix, Ariz. (current edition).
14. *Linear Data Book*, National Semiconductor Corp., Santa Clara, Calif. (current edition).
15. S. K. Mitra (ed.), *Active Inductorless Filters*, IEEE Press, New York, 1971.

16.  S. S. Haykin, *Active Network Theory*, Addison-Wesley Publishing Co., Reading, Mass., 1970.

17.  R. P. Sallen and E. L. Key, "A practical method of designing RC active filters," *IRE Transactions on Circuit Theory*, CT-2, pp. 74–85, March 1955.

18.  J. Tow, "Design formulas for active RC filters using operational amplifier biquad," *Electron. Letters*, pp. 339–341, July 24, 1969.

19.  M. E. Van Valkenburg, *Introduction to Modern Network Synthesis*, John Wiley and Sons, Inc., New York, 1960.

20.  L. Weinberg, *Network Analysis and Synthesis*, McGraw-Hill Book Co., New York, 1962 (reprinted by Robert E. Krieger Publishing Co., Melbourne, Fla., 1978).

21.  D. E. Johnson, *Introduction to Filter Theory*, Prentice-Hall, Inc., Englewood Cliffs, N.J., 1976.

22.  D. E. Johnson and J. L. Hilburn, *Rapid Practical Designs of Active Filters*, John Wiley & Sons, Inc., New York, 1975.

23.  A. Papoulis, "On the approximation problem in filter design," *IRE National Convention Record*, vol. 5, pt. 2, pp. 175–185, 1957.

24.  H. Ruston and J. Bordogna, *Electric Networks: Functions, Filters, Analysis*, McGraw-Hill Book Co., New York, 1966.

25.  W. J. Kerwin and L. P. Huelsman, "The design of high performance active RC band-pass filters," *IEEE International Convention Record*, vol. 14, pt. 10, pp. 74–80, 1960.

26.  L. P. Huelsman, *Theory and Design of Active RC Circuits*, McGraw-Hill Book Co., New York, 1968.

27.  P. R. Geffe, *Simplified Modern Filter Design*, Hayden Book Co., Inc., New York, 1963.

28.  L. P. Huelsman, *Active Filters: Lumped, Distributed, Integrated, Digital, and Parametric*, McGraw-Hill Book Co., New York, 1970.

29.  T. Deliyannis, "RC active allpass sections," *Electron Letters*, vol. 5, pp. 59–60, February 1969.

30.  L. Storch, "Synthesis of constant-time-delay ladder networks using Bessel polynomials," *Proceedings of the IRE*, vol. 42, no. 11, pp. 1666–1675, November 1954.

# INDEX

## About the Authors

JOHN L. HILBURN, Ph.D., P.E., is president of Microcomputer Systems, Inc., a firm specializing in the design and fabrication of microcomputer-based industrial instrumentation. A former professor of electrical engineering at Louisiana State University, Dr. Hilburn is the author of 12 books and numerous technical papers and a member of IEEE.

DAVID E. JOHNSON, Ph.D., P.E., is professor of electrical engineering at Louisiana State University, where he has been since 1962. A member of IEEE, Dr. Johnson is the author of 17 books and some 40 technical papers.